Organic Public Engagement

Adam S. Lerner • Pat J. Gehrke

Organic Public Engagement

How Ecological Thinking Transforms Public
Engagement with Science

Adam S. Lerner
University of South Carolina
Columbia, South Carolina, USA

Pat J. Gehrke
University of South Carolina
Columbia, South Carolina, USA

ISBN 978-3-319-64396-0   ISBN 978-3-319-64397-7   (eBook)
https://doi.org/10.1007/978-3-319-64397-7

Library of Congress Control Number: 2017953102

Cover illustration: © JDawnInk

Printed on acid-free paper

This Palgrave Macmillan imprint is published by Springer Nature
The registered company is Springer International Publishing AG
The registered company address is: Gewerbestrasse 11, 6330 Cham, Switzerland

ACKNOWLEDGMENTS

Producing any book is never the work of just one or even two people, but reflects the contributions of countless colleagues, friends, and supporters. We are grateful to have benefited from the support of the National Science Foundation, which provided the resources necessary to conduct the fieldwork that tested and refined the organic public engagement methodology.[1] The editors and reviewers at *Public Understanding of Science* were also vital in developing the first iteration of the methodology. While this book is a dramatic expansion on that article and further improvement on the methodology, we also found a few small sections of that original essay were especially useful at key points in the book. A highly attentive reader familiar with the original article will find a duplicated sentence here or there, scattered throughout the book. We are grateful that the publication agreement with the journal and its publisher (SAGE) permitted such use.

A number of colleagues were also essential to the maturation of this project. We are grateful to David Berube at North Carolina State University for inviting one of the authors (Pat Gehrke) to develop and run the public engagement events we discuss in Chaps. 6 and 7, as well as for his input on the earliest articulations of the organic public engagement methodology. We are also indebted to the work of two graduate research assistants who worked as coders on that study: Matthew Boedy and Jonathan Maricle. Pat would also like to thank John Stone, who provided many excellent conversations during his time with Michigan State University's Center for the Study of Standards in Society.

Finally, none of our work would be possible without the support of our department, college, and university. The University of South Carolina has been supportive of this work, and we especially thank the College of Arts and Sciences for providing additional funding that brought this project through its final months of development.

NOTE

1. National Science Foundation Grant: Nanotechnology Interdisciplinary Research Team in Intuitive Toxicology, #06-595.

CONTENTS

CHAPTER 1

Introduction

In a large meeting room of the public library in a mid-western town, about sixty people have gathered to hear about invisible and unlabeled risks in cosmetics and sunscreens. On a table at the front of the room is a collection of popular name-brand products gathered by three members of a local consumer advocacy group. They are the organizers of this event and have set its agenda, decided its format, found its location, created the content, and done all the publicity. Almost everyone in this room knows each other; they have gathered for similar meetings in the past about issues like local water quality and concerns about food additives. In the back of the room are two tables, one with snacks, water, tea, and coffee. The other holds fliers, handouts, a clipboard with a petition, and a stack of notecards pre-addressed to local state and federal legislators. Tucked away in one of the back corners is a tripod with a small video camera, recording the entire event. It belongs to a man seated in the back row. He does not speak during the event, but is attentive and taking copious notes.

A month earlier that same man and his camera were in the back of a large lecture hall at a community college in a rural southern town. He watched and took notes as two speakers were grilled by an audience of nearly a hundred people. Both speakers were billed as experts on the risks of nanomaterials, tiny engineered molecules that have unique and often astounding properties. The participants were flummoxing both speakers with questions about imbuing particles with spin, the rate of decay of a charge, and how complex technical limitations were overcome to produce these molecules. The speakers had not expected these kinds of

© The Author(s) 2018
A.S. Lerner, P.J. Gehrke, *Organic Public Engagement*,
https://doi.org/10.1007/978-3-319-64397-7_1

questions. After all, this was a public engagement event organized by the adult continuing education branch of the local community college. It just so happens that the community college is not far from a national research laboratory and their adult continuing education program consistently attracts scientists, physicists, and engineers who have retired in the area. As regular participants in the community college programs, many of the audience members knew each other and had established expectations and norms that made grilling these presenters acceptable and expected.

Each of these scenarios depicts a public engaging with science and technology. The man with the camera taking notes was one of the authors of this volume (Pat Gehrke), studying how these publics engage with science. Three years before these events, he had been working with a team of scholars on a grant proposal to study public concerns about the risks of nanotechnologies. His contribution was to articulate the public engagement portion of the grant, including its methodology and underlying philosophy. The approach to public engagement he proposed, which was used for these events and nine more like them, differed from how most researchers conducted public engagement events. Those differences came from research on what constitutes a public, how human communication functions in relationship to its ecology, and what it means to participate and engage. At the time, the idea for this public engagement methodology was largely philosophical and theoretical, but so much of the research in communication, psychology, and related fields supported the need for a shift in this direction that he was confident he was on the right track. He just needed funding for the fieldwork to test and refine the method.

After the National Science Foundation funded the grant, he found himself presenting this methodology at an NSF meeting of primary investigators, all of whom were working on public engagement with science or public understanding of science. He knew his approach was unconventional, but so much of the research and theoretical literature supported the new methodology he was confident enough to share the argument with his peers. The response was not simply mixed, but polarized. Half or slightly more of the attendees were vehemently opposed to the ideas, staunchly defending methods that, by his reading of the research, were simply not designed to investigate public understanding or engagement with science, much less support the claims researchers sought to make. Slightly less than half were equally vocal and firm in their agreement with his arguments and the need to rectify core deficiencies in the most commonly used models of public engagement. Clearly he was on to something, but as of yet, neither he nor his colleagues could fully articulate their positions, and the method was untested and unrefined.

Over the following five years, through numerous presentations and discussions with colleagues, eleven public engagement events deploying various refinements of the methodology, and dozens of drafts, the methodology reached maturity. This method is called "organic public engagement" and the first short version of its articulation was published in *Public Understanding of Science* in 2014. This book is a more complete and extended argument for the need to move our public engagement with science toward such organic methodologies. As the following chapters document, organic public engagement differs from most of the dominant models of public engagement by paying closer attention to actually existing publics and how they behave within their existing ecologies. As its name implies, organic public engagement seeks to work with existing local conditions and structures to better understand actually existing publics and create more sustainable engagement outcomes. Likewise, the name implies that many other forms of public engagement are something other than organic, perhaps even artificial. We go to great lengths to provide both concrete empirical research and strong theoretical foundations to evidence the artificiality of dominant methods of public engagement with science while also documenting the significant costs of such artificiality. Part of that argument is based on a disconnect between the kinds of "publics" or groups of people who are gathered by artificial public engagement researchers and how people gather to form publics without the intervention of engagement experts and researchers. We argue that unless we can effectively engage these actually existing publics, as they exist in their given ecologies, then public engagement with science will be too artificial to generate useful research or sustainable public benefits. This artificiality is generated in large part by a disconnect between artificial public engagement and a century of consistent and compelling research on the role of ecology in human communication and behavior.

Our purpose in writing this book, however, is not merely to level a critique against artificial public engagement methods. Our goal is to provide a practicable alternative that is grounded in current empirical research, historical perspective, and sound theoretical foundations. In the chapters that follow, we have endeavored to provide the reader with a broad introduction to the history and principles of research in ecological thinking and ecological validity, with special attention to how ecology impacts both human behavior and the validity of social scientific research. That research, combined with work by scholars of communication and political science, yields a more robust understanding of what constitutes a public and what it might mean to engage actually existing publics. In our view, once one is familiar with the research on ecology, ecological validity, human communication, and publics, the critique of artificial public engagement becomes rather obvious and quite damning. What is less obvious and make us hopeful is that the same research leads naturally toward an alternative: the general methodology of organic public engagement and specific methods both for conducting research and building theory from that research.

Chapter 2 introduces the reader to ecological thinking, beginning with early work on ecologies in botany, biology, and geology in the nineteenth century. We then move to the development of theories of ecology in human and social sciences in the twentieth century, with particular attention to anthropology, sociology, and psychology. That foundation then permits the introduction of the standard of ecological validity (first introduced in the 1940s) and its application to the study of deliberation and communication. From this rich history and robust body of research, we articulate five standards for ecological validity in public engagement with science.

Chapter 3 advances our critique of artificial public engagement with science by applying the five criteria from Chap. 2 (and the research supporting each) to common methods of public engagement. We begin by considering survey studies, media studies, science fairs and cafés, museums, and digital engagements, providing recommendations for each based on ecological thinking. The bulk of the chapter, however, focuses on deliberative engagements such as consensus conferences, citizen panels, citizen juries, and deliberative polls. We argue that each of these, as currently conceived and practiced, has serious deficiencies in their ecological validity. These deficiencies are common, we contend, because these deliberative methods are built upon certain political and philosophical commitments that produce both cynical and idealist attitudes toward the publics they study.

Chapter 4 details these political and philosophical commitments that drive artificial public engagement with science. We trace their origins through the history of political philosophy, with some attention to Immanuel Kant, Walter Lippmann, John Dewey, John Rawls, and Jürgen Habermas (the last two being especially important to current theorists supporting artificial public engagement methods). We then contrast their assumptions with more recent research on publics and public communication, especially studies of counterpublics and vernacular rhetoric. We argue that this work produces a more practical, balanced, and sound basis for public engagement with science.

Modern theories of publics and counterpublics, combined with studies of vernacular rhetoric, offer us a richer approach to considering what constitutes publics, how they form and change, and how they can be studied.

Chapter 5 then confronts one of the thorniest problems in public engagement with science (and to a lesser degree, public engagement in general): the demarcation between experts and nonexperts. We begin by noting the long history of the problem in philosophers' challenge to differentiate types of knowledge. We then move specifically to the issue of scientific knowledge, working through the theories of Karl Popper and Thomas Kuhn. As we are ultimately concerned with how these philosophical and theoretical issues impact actually existing publics and practices of engagement with science, we turn next to how the demarcation problem has played out in political contexts, especially in the courts. This leads us to the call for a "third wave" of science studies, the utility of counterpublics and vernacular rhetoric research in confronting the demarcation problem, and how methods drawn from human communication scholars can aid in this research. We also note here the irony of much public engagement with science, wherein the "expert" priesthood of science is defrocked and asked to stand in equal conversation with "nonexpert" lay citizens, but self-styled engagement experts claim a special status as the holders of the truth of democratic politics.

Having concluded the foundational research and critique that drives our turn toward organic public engagement, Chap. 6 defines and lays out the methodology in some detail. Organic public engagement methodology draws from a variety of fields, informed by the theories and research covered in the previous four chapters. We provide researchers and practitioners enough foundation to conduct organic public engagement with science events and walk them through how one study implemented the methodology. A reader solely interested in implementing organic public engagement with science could jump here, but we caution from our own experience that the method is more easily and fruitfully deployed once one has at least an introduction to theories of ecology, the principles of ecological validity, research on counterpublics and vernacular rhetoric, and the demarcation problem.

Chapter 7 concludes the book by discussing how organic public engagement events can be used to generate meaningful research outcomes and build theories. We believe that organic public engagement is especially well suited to building middle-range theories or heuristics. Such theories and heuristics, we argue, are more actionable and meaningful outcomes for public engagement with science than the more "grand" theories science sometimes seeks. By drawing from research in multi-sited ethnography and the methods of grounded theory, we argue that one can collect outcomes from multiple organic public engagement events to make more useful and robust theories and heuristics.

As with any methodology, organic public engagement should not be considered complete or finished. In fact, one of the important dimensions of organic public engagement as a methodology is that it should always be adapting and growing to meet the conditions in which it is deployed. We hope that the reader finds in this book both sufficient reason to take up the challenge of conducting organic public engagement with science events and also the practical tools for doing so.

CHAPTER 2

Ecological Thinking in Science and Public

In the late nineteenth and early twentieth centuries, Yellowstone National Park implemented a rigorous wolf control program, all but eliminating their presence. Surprisingly, the elimination of wolves also brought a marked decline in the population of northern range beavers as well as fewer aspen, cottonwood, and willow trees in the northeastern sections of the park. In 1995 the park service reintroduced wolves into the ecosystem, and by the early twenty-first century, the population of these deciduous woody trees and their beaver neighbors had once again flourished. Herein lies the mystery: how is it that a decline in the wolf population results in fewer beavers and trees? What sort of relationship do wolves, trees, and beavers have?

The composition of this relationship turns out to be both simple and unexpected. Without wolves in the park, elk no longer needed to escape from predators and therefore remained stationary, overgrazing the woody trees. In turn, without lush woody trees in the park, beavers had no means to construct shelters and moved on to more fertile areas. This chain of unexpected cause and effect, called a "trophic cascade," provides a picture of what we mean when we say "ecology." Conceptualizing the connection between inhabitants (both the human and nonhuman varieties) and habitats constitutes what we call "ecological thinking," which is the study and consideration of how organic and inorganic elements of life interact.

This chapter traces a broad history of ecological thinking, starting with the early proto-ecological natural scientists in the eighteenth and nineteenth centuries. After laying out some of these early conceptions of ecology, we then turn to the social sciences and humanities, exploring how ecological

© The Author(s) 2018
A.S. Lerner, P.J. Gehrke, *Organic Public Engagement*,
https://doi.org/10.1007/978-3-319-64397-7_2

thinking influenced sociology, anthropology, psychology, ethnography, and philosophy. From this foundation, we suggest that ecological thinking can be used to revisit the concept of ecological validity, originally developed in the mid-twentieth century by psychologists Egon Brunswik and Kurt Lewin. This chapter concludes by explaining how ecological validity and ecological thinking in general provide public engagement with science both concrete guidelines for better research and heuristics for evaluating the strengths and weaknesses of competing engagement methods.

PROTO-ECOLOGY IN NATURAL SCIENCE: EARLY DEVELOPMENTS OF ECOLOGICAL THINKING

The proto-ecological thinkers of the mid-eighteenth and early nineteenth centuries are bound together by their agreement with Aristotle's argument in Book 1 of the *Physics*: all of nature is intended for the use of human beings. This notion is, of course, echoed in Genesis, where God gives domain over nature to Adam in the Garden of Eden. Thus, proto-ecological thinkers in the natural sciences were responding to cultural, religious, and historical demands to conceptualize nature as humanity's rightful possession. In doing so, they demonstrated an understanding of the connection between humans and nonhumans, but their understanding functioned only hierarchically, positioning humanity as nature's master.

Carl Linnaeus

Carl Linnaeus, the eighteenth century Swiss botanist, demonstrates this kind of proto-ecological thinking in many of his published texts. Due to his analogical approach to plant reproduction and his theory of an economics of nature, Linnaeus stands as one of the earliest examples of an ecological thinker. Known for his immense taxonomy, Linnaeus's most recognizable contribution to botany may be his theory of plant reproduction in his 1737 *Genera Plantarum*. In addition, his classificatory structure in *Systema Plantarum* (1753) and *Systema Naturae* (1758) persist as foundations for the modern nomenclature of plants, macroscopic animals, viruses, bacteria, crops, garden plants, and genetically engineered organisms (Koerner 1999, p. 16).

Despite these contributions, the connection between Linnaeus and ecological thinking is seldom made. Michel Foucault (1970) hinted at this connection when he noted that Linnaeus's classificatory system in *Philosophia*

Botanica adheres to four variables: "the form of the elements, the quantity of those elements, the manner in which they are distributed in space with relation to each other, and the relative magnitude of each element" (p. 134). This third variable, how elements of an organism are interrelated, constitutes an example of ecological thinking, albeit only analogically.

A more substantial example appears in Linnaeus's obscure 1775 text *Oeconomia Naturae* or *The Economics of Nature*, which stands as a major development in what would become the modern field of ecology (Dove 2015, p. 239). This often-overlooked text by Linnaeus wove an understanding of human economic concerns with his knowledge of agriculture and botany. As Michael Dove (2015) explains, Linnaeus believed that humans were part of "nature's economy," which was tailored by a Creator to help humans thrive. This perspective is apparent in Linnaeus's remarks on the importance of species differentiation and soil fertility, where he underscores the important role that humans play in maintaining and preserving the natural world.

Alternatively, Linnaeus's then-controversial stance on the practice of swiddening ("slash-and-burn" agriculture) shows how deeply this proto-ecologist understood the sometimes-tenuous relationship between humanity and nature. Both condemning the practice (for its detrimental ecological impact) and understanding its importance to the Laplanders, Linnaeus was "professionally and theoretically at home in an anthropogenic landscape" (Dove 2015, p. 234). Linnaeus grasped that while "slash-and-burn" agricultural methods were harmful to the environment, they also offered the indigenous Finno-Ugric people a method of sustaining their populations.

James Hutton

In the field of geology, James Hutton's (1788) *Theory of the Earth* is another early example of ecological thinking. Hutton conceived of the earth and its inhabitants as a living system, but today, he is probably best known for his claim that the world was more than 6,000 years old, a contentious claim at the time given its direct contradiction of the book of Genesis. Truly the forefather of modern geology, his systematic theory of the earth also stands as the precursor to much of our knowledge about plate tectonics, and many of his theories influenced Charles Darwin. Much like Linnaeus, Hutton saw the earth as a system constructed by a divine Creator for the purpose of human inhabitance. Take, for instance, this short passage in *Theory of the Earth*: "The globe of this earth is evidently made for man... he alone can make the knowledge of this system a source of pleasure and the means of happiness" (Hutton 1788, p. 217). While Linnaeus saw the relationship between earth and humans as economic, Hutton saw it as analytical or even aesthetic.

For Hutton, the divine Creator constructed the earth much like a puzzle for humans to solve: "It is with pleasure that he observes order and regularity in the works of nature, instead of being disgusted with disorder and confusion; and he is made happy from the appearance of wisdom and benevolence in the design, instead of being left to suspect in the Author of nature, any of that imperfection which he finds in himself" (Hutton 1788, pp. 287–288). Blending a view that combines both the Enlightenment's dedication to empirical science with a deistic conception of a Creator, Hutton articulates a basic understanding of ecology. However, this understanding is not one centered around human's physical, agricultural, or environmental connection with nature. Rather it is concerned with human's connection with nature via the intellect. For Hutton, the connection between the chaos of nature and the intellect of humans is one of pleasurable reflection and contemplation. By making sense of the world, we are in turn part of the world.

Jean-Baptiste Lamarck

Jean-Baptiste Lamarck is more controversial than either Linnaeus or Hutton, but his contribution to ecological thinking should not be overlooked. His most famous innovation in natural science is probably his distinction between vertebrates and invertebrates in 1797, which was the catalyst for his more contentious system of transformism (Corsi 1988, p. 65). In its strongest form, "Lamarckism" posits both an increasing inclination of creatures toward complexity and a "use and disuse" doctrine, where characteristics are developed and passed down to future generations based upon the use and disuse of these characteristics in a given environment. For example, in this system, a giraffe repeatedly reaching for leaves in treetops "stretches out" its neck over time. A longer neck then becomes a feature that is passed down to its offspring.

While Lamarck's evolutionary theories have largely been debunked (and often misrepresented), his influence as an ecological thinker is best understood in relation to a new field of research called "epigenetics." First coined by Conrad Waddington in 1942, epigenetics is a field dedicated to describing events that could not be explained by purely genetic principles (*epi-* meaning "outside"). A first wave of research in this field came in the 1980s, with advancements in genetics. Unlike genetics, however, epigenetics is interested in how the environment affects how cells express their genes, even though their underlying genetics remain unchanged. These normal processes, to put it simply, alter both the appearance and structure of the gene, changing the way that it behaves. To put it in another sense, this process can "silence" genes, preventing them from copying their genetic code. In the 1980s, the first human disease associated with epigenetics was cancer (Feinberg and Vogelstein 1983, p. 91).

More recent epigenetic research in the fields of behavioral science and criminology has somewhat vindicated Lamarck's basic theories (DeLisi and Vaughn 2015, p. 608).[1] Lamarck's elementary principles, while flawed, have been reconceptualized in modern biological research techniques and used to describe how the environment can affect humans at the cellular level. While this research needs further validation, it supports the argument for an ecological basis of our existence.

Alexander von Humboldt

Alexander von Humboldt, the famous German naturalist, came to think in ecological terms after two vital observations. The first occurred at Lake Valencia around 1800, where, only a few decades earlier, forests had shielded the soil from the sun and diminished the evaporation of rain waters. Without these trees, Humboldt noticed how rains would inundate the loosened unprotected soil, damaging the banks of the lake. Humboldt had already noted the effect that logging had in the Fichtel Mountains near Bayreuth a few decades earlier, and his reports were filled with suggestions for reducing the need for lumber (Wulf 2015, p. 58). Now, however, he understood the effects of deforestation in a new light; extrapolating from his findings at Lake Valencia he warned that the agricultural techniques of his time could create irreparable and long-lasting effects on the environment.

His second observation happened only a few weeks after the first, deep in the Orinoco rainforest, where he observed Spanish monks using oil harvested from turtle eggs. Their simple act of over-harvesting the eggs had a significant impact on the local turtle population. Earlier in his career he had made the same observation at the Venezuelan coast about the effects of pearl harvesting on oyster stocks. Humboldt turned away from the Aristotelian view that nature was intended for humans, a belief that the Bible, Francis Bacon, Rene Descartes, Linnaeus, and Hutton all had taken for granted. In turn, Humboldt refuted the belief that humans should conquer the land and control it, a view popularized by the French naturalist Comte de Buffon (who, coincidentally, influenced Lamarck). Humboldt had spoken the first whispers of our contemporary conversation on humanity's environmental impact.

Ernst Haeckel

The interrelation between human and environments is both give and take. Not only are humans subject to the conditions of their environment, but they also produce that environment. Our home shapes us as we shape it. This realization was coming to fruition just as Ernst Haeckel, sometimes called the "German Darwin," coined the term "ecology" and successfully blended cell theory with evolutionary theory. His theory of how multicellular organisms form from colonies of single-cell organisms is still relevant, as is his promotion of an evolutionary morphology, which organized and arranged organisms according to a hierarchy of increasingly complex levels of individuation (Reynolds 2008, p. 123).

Utilizing Darwin's theory that creatures live and are implicated in complicated networks of inorganic and organic environments, Haeckel coined the term ecology in 1866 in his *Generelle Morphologie*, and defined it as the "science of the mutual relationships of organisms to one another" (Haeckel 1866, quoted in Richards 2008, p. 144). Sadly, Haeckel's views are mostly known today for their ties to Nazi conceptions of *volk* (and by implication, practices of eugenics). However, Haeckel's originality and contribution to the popularization of evolution as well as his development of the field of ecology left an important mark on the history of ecological thinking.

This cursory history of proto-ecological thinking in the eighteenth and early nineteenth centuries reveals two primary schools of thought. Linnaeus, Hutton, and Lamarck all demonstrated a simplistic and implied form of ecological thinking, bound by a hierarchical view that humans dominate over the earth as both keepers and tamers. However, Humboldt, Haeckel, and Darwin all broke from this early mode of ecological thinking, and began to conceive of ecology as a relationship of mutuality. What binds their conception of ecology is an emphasis on connectivity, interrelation, and reciprocity. Their contributions to the natural sciences are fundamental and important, and create the conditions of possibility for more systematic approaches to conceptualizing what ecologies are, what they do, and how they function.

Human Ecologies: Ecological Methodology in Sociology, Anthropology, Ethnography, and Psychology

Certainly, much more could be said about the history of ecology in distinct fields of natural science: in the field of botany, for instance, we could point to August Grisebach's work in the mid-nineteenth century, or early botanical ecologists such as Isaac Biberg and Andreas Hedenberg in the mid-eighteenth century. However, the goal of this historical overview was not to give an extended treatment of ecology's foundations. Rather, the goal was to establish a rough baseline from which offshoots and extensions of basic ecological principles derive.

As we pivot to how ecological thinking is taken up by sociologists, anthropologists, psychologists, ethnographic researchers, and science studies scholars, we treat ecology in terms of movements or moments in these fields of human inquiry. Part of this pivot is a transition from ecological *thinking* to ecological *methodology*. In the natural sciences, Frederic Clements's 1905 volume *Research Methods in Ecology* is one of the first manuals devised for what we now know as ecological sciences, and it stands as a decisive contribution to this burgeoning field of study. His work on plant communities and ecologies was widely influential until the middle of the twentieth century and experienced something of a revival near the beginning of the twenty-first. However, Clements development of a systematic ecological methodology was also replicated by social scientists and humanists in the twentieth century. Starting with the Chicago School of sociology, we will trace the various ecological developments that appear in these fields. These developments are derived from same conceptual space of the proto-ecologists, but focus more on human behaviors and take on a more systematic character than the efforts of eighteenth or nineteenth century natural philosophers.

The Beginning of Human Ecology and the Chicago School

Sociologist Roderick McKenzie aptly summarized the state of ecological studies in the wake of Clements's work when he lamented in 1924 that there had "developed no science of human ecology which is comparable in precision of observation or in method of analysis with the recent sciences of plant and animal ecology" (McKenzie 1924, p. 287). To remedy this situation, the young McKenzie, along with his mentor Robert Park and colleague Ernest Burgess, developed a methodological approach to human ecological studies. Their best-known work is the seminal *The City*, which deployed what McKenzie called the method of "human ecological study." Their collected efforts are now referred to as the "Chicago School" of sociology.

Combining an analysis of the physical parameters of institutional structures with a sociological analysis, McKenzie and his colleagues produced an ecological movement in sociology. For McKenzie, ecology as it existed in the fields of biology and botany failed to capture the multiple levels of complexity that we find in human interactions. McKenzie openly admitted in 1924 that studies had been done about the biological, social, or economic impacts of competition and selection on humans, but all of these studies failed to account for the distributive and spatial aspects of these processes and, therefore, had failed to account for human ecology (McKenzie 1924, p. 288). In an effort to correct this shortcoming, he developed the definition of human ecology as the "study of the spatial and temporal relations of human beings as affected by the selective, distributive, and accommodative forces of the environment" (McKenzie 1924, p. 288).

Both Burgess and Park were central contributors to the development of this field of study, but McKenzie's theoretical work in the mid-1920s was the foundation needed for this emerging perspective to gain a foothold. In his writings, McKenzie differentiates plant and animal ecology from human ecology by humans' sophisticated ability to alter the environment to suit their purposes. Combined with his emphasis on spatial distributions, McKenzie's basic presupposition was that humans' physical engagement with the environment (natural and social) is both precipitated by technological advancements that affect space/time (automobiles and telephones) and understandable in terms of social groupings (family and coworkers).

This mode of ecological thinking contributes to the kinds of questions found in *The City*: how does city planning affect how neighborhoods are created? How do those neighborhoods maintain a common set of customs and mores? How does living in close proximity to people of other classes or races (or segregation from other people) change the sentiments and customs of a neighborhood? How does access to urban transportation affect the jobs, trades, and class structures that we see in the city? These questions are as relevant today as they were in 1925, and each relies on an ecological perspective, where humans' relationships to one another and the institutional lives they inhabit are affected by spatial patterns, flows, organizations, and distributions.

McKenzie utilized this human ecological methodology to great effect. Applying this approach, he anticipated modern globalization and the increasing standardization of material culture, which he credits to the shockingly rapid development of new transportation and communication technologies in the 1920s (MacDonald 2011, p. 274). Despite McKenzie's seeming prescience, the Chicago School's analysis of urban life in Chicago is also noticeably dated. Many of the observations these researchers made are peculiar to their historical and cultural ecologies, so to speak. In addition, many of the authors' claims are not supported by solid evidence, and they often assume the validity of suspect theories. Alternatively, at points their tone shifts from a descriptive to prescriptive mode of address. For instance, Park and his coauthors praise the temperance movement of the early twentieth century, calling its effects "highly picturesque" (Park et al. 1925, p. 32). Regardless of the potential deficiencies in their analysis, their work still stands as a sophisticated example of how ecological thinking transitioned from the study of plants and animals to the study of humans.

Although methodologically sophisticated, their work was not easily accepted by their peers. On the contrary, McKenzie's research was met with resistance from other sociologists at the time, who claimed that his ecological method "focused on habitat to the neglect of humans," and that this new form of human ecology "neglects social-psychological aspects of community and cultural factors, and subscribes to a form of biological determinism" (MacDonald 2011, p. 266). As Dennis MacDonald (2011) suggests, however, most of the activities described by McKenzie's ecological methodology retain a cultural core while addressing physical aspects of social life (p. 268). While McKenzie does tend toward explaining phenomena in terms of the physical institutions that humans inhabit, his theory does not displace or ignore the role that cultural or social factors play in city formations. The problem of determinism is one that will reemerge as we turn toward our next site of ecological methodology, the study of cultural ecology in anthropology.

Cultural Ecology in Anthropology: Julian Steward and Environmental Determinism

Despite the historical developments in cultural ecology over the years, Julian Steward's work stands as a vital example for understanding how ecological thinking became a research methodology in anthropology. His 1955 classic, *Theory of Culture Change*, conceived the cultural ecological approach. Ecological thinking had made a mark in anthropology before Steward's formulation of cultural ecology, but these early studies fell prey to environmental reductionism. For instance, studies by Lewis Morgan, Edward Tylor, and others in the nineteenth century developed a model of cultural evolution that progressed in discrete, fixed stages. For instance, Morgan constructed a seven-stage evolutionary model developed around categories devised by Karl Marx and Friedrich Engels. These models simplified and reduced cultures to a scale of evolution, allowing anthropologists to assert that certain cultures were more or less "evolved" than other cultures.

These approaches inevitably failed when researchers attempted to fit a growing amount of data into their preconceived categories. Increasingly detailed accounts of supposedly "unevolved" cultures were incompatible with the neat, tidy, and one-dimensional analyses that cultural evolutionists developed. Reactions to this approach were so strong, in fact, that it caused researchers who opposed this methodology to band together, which in turn was a major catalyst for the establishment of anthropology as a distinct academic discipline (Orlove 1980, p. 236). Out of this academic commotion, Julian Steward emerged in the middle of the twentieth century with his method of cultural ecology.

Largely an heir to the British school of social anthropology, Steward's contact with the famous geographer Carl Sauer pushed him to better understand the effects of environment on culture (Orlove 1980, p. 246). For Steward, the method of cultural ecology "entails the study of the relation between certain features of the environment and certain traits of the culture possessed by the sets of people living in that environment" (Orlove 1980, p. 235). Within this environment, Steward was particularly interested in technologies, economics, demography, and social arrangements, but he stressed that the environment only affects certain elements of a culture, elements he would call a "cultural core." Elements of a culture outside of this "core" were subject to more traditional modes of development, and elements within it were related to complex environmental factors.

Steward's development of cultural ecology was also, in some respects, indebted to the work of McKenzie, Park, and Burgess in the Chicago School. However, per Steward (1972), their human ecology lacked the ability to "throw any light on world-wide ecological urban adaptations, for in other cultures and periods city zoning followed very different culturally prescribed principles" (p. 33). This posed a fundamental scientific problem for Steward, as he wrote in *Theory of Culture Change*:

Is the objective to find universal laws or processes, or is it to explain special phenomena? In biology, the law of evolution and the auxiliary principles of ecology are applicable to all webs of life regardless of the species and physical environments involved. In social science studies, there is a similar effort to discover universal processes of cultural change. But such processes cannot be conceptualized in biological terms...Until the processes of cultural ecology are understood in the many particulars exemplified by different cultures in different parts of the world a formulation of universal processes will be impossible. (Steward 1972, pp. 33–34)

A tall task, but in seeking a more generalizable ecological methodology, Steward realized this problem had to be confronted. One contribution to building up an archive of these "particulars exemplified by different cultures" was his study of Native American tribes in the American Southwest (more specifically the western Puebloan people). In this study, he argued that the "initial development of patrilineal bands and of localized, exogamous matrilineal lineages is a simple cultural ecological adjustment" (Steward 1972, p. 171).

Steward's study of the Puebloan people demonstrated his ecological methodology in five steps. First, Steward theorized that if a society is horticultural (which requires arable land), it might consist of a localized, exogamous, and possibly landowning people. Second, if these horticulturalists have an increased food supply or other factors that would lead to a denser population, either there are larger social groups occupying one territory, or separate groups which occupy less territory, or multi-lineage villages in which several previously localized groups live together. Third, multi-lineage villages come to exist in virtue of movements, war, or other factors that might dislodge bands or lineages from their villages, causing them to comingle in the same locality. Fourth, while it is not necessary that these comingled groups become a cohesive unit, they will do so if they share a common name, ceremonies, or other factors that produce a sense of solidarity through generations (even if these lineages are no longer traceable). Fifth, over the course of these changes "political autonomy passes from the localized lineage to a larger group, and a higher level of sociocultural integration appears – the Pueblo village, the Yuman nation" (Steward 1972, p. 171). Beginning from farming arable land and ending with the development of cultural units like the village or the nation, Steward utilized an ecological methodology to great effect.

Each step of his logic demonstrates an ecological methodology, which seamlessly weaves an understanding of Pueblo culture with the conditions of the surrounding environment. Steward is noticeably careful in his analysis, however. He makes certain to put forward his theory as preliminary, and articulates evidence for his argument in a multilinear, multicausal framework. Multi-lineage villages may emerge due to the distribution of arable land, but it is not the *only* reason multi-lineage villages may emerge, nor is it clear that a multi-lineage village will appear at all due to particular environmental factors. Steward noticed and attempted to avoid the problem of environmental determinism.

Acknowledgment of the potential pitfalls of environmental determinism is necessary if researchers wish to engage with an ecological methodology, and Steward's careful consideration of the causal mechanisms which lead to cultural developments in the Puebloan people serves and an excellent guiding example. While the environment is certainly a part of our interrelationships with one another and ourselves, using the environment as the only mode in which humans are understood is ironically, highly unecological. Using Jared Diamond's popular 1997 work *Guns, Germs, and Steel* as central object of study, Gabriel Judkins et al. (2008) chart different levels of adherence to environmental determinism in human-environment research from the late nineteenth century to modern day. There is some amount of gradation in how much or how little a specific theory adheres to environmental determinism; however, adherence to this theory poses a distinct problem for ecologists of all stripes. By placing the environment as the determining factor in human development, the proto-ecological problem of hierarchy is simply reversed. Instead of humans being at the center of all things, nature is instead given a privileged position. Judkins, Smith, and Keys conclude that the history of ecological research is fairly divided in its attention between societal, familial, and environmental factors (Judkins et al. 2008, p. 20). Ecological determinism is a problematic hurdle, but there is little evidence that this is a widely held or popular position.

Environmental Psychology: Barker's Oskaloosa Study

Just as in the fields of sociology and anthropology, ecological methods of investigation also emerged in the field of psychology in the twentieth century. In general, the development of an ecological methodology in psychology occurred by means of three major contributions: first is the work of Brunswik and Lewin in the 1940s and 1950s, who paid sustained attention to the potential that ecological methodologies might afford psychological studies. Next, the work of Barker at the Midwest Psychological Field station in the 1960s paved the way for a more comprehensive methodological apparatus grounded in ecological thinking. Finally, the work of James Gibson on the ecology of perception in the 1970s expanded the potential applications of an ecological methodology to the visual field. There are, of course, more contributions to this lineage of ecological methodology found in psychology (the work of Ulric Neisser on ecological cognition, for instance), but these three authors are central for understanding how ecological methodologies have developed in the field of psychology.

The work of Brunswik, Lewin, and Gibson are fundamental to the final section of this chapter, which (re)introduces the concept of ecological validity, so we reserve these authors and their contributions for that discussion. However, we should briefly consider the work of Barker, who made important contributions to understanding how ecological methodologies contribute to psychology. As he points out in his 1968 work *Ecological Psychology*, Barker (1968) sees the tendency for psychologists to utilize lab settings as a self-imposed limitation on psychological research (p. 1). While psychologists at the time had a good sense of how people behave under the conditions of experiments and clinical procedures, they knew little about how these conditions are distributed and occur outside of the laboratory setting, or how people realistically react to these conditions. Barker's (1968) solution to this problem was to go out into the field to record and analyze "the psychologist-free environment of behavior" (p. 4).

Barker conducted his analyses from the late 1940s through the early 1970s at the Midwest Psychological Field Station in Oskaloosa, Kansas, a project funded by the University of Kansas. He and his colleagues generated a substantial amount of empirical data during this period, and one prime example is *One Boy's Day*. In this work, Barker recorded the day-to-day activities of a seven-year-old boy called "Raymond," starting from when the boy woke to when he went to bed. For instance, Chap. 3 covers a period in Raymond's class from 8:53am to 9:24am. The best way to describe Barker's account is by quoting his analysis directly. The following events took place from 8:55am to 8:56am, and immediately follow Raymond's attempt to draw a steam boat with his classmates:

> Radiating his arms loosely from the shoulder, he whirled them round and round. He broke into a jog with short choppy steps. He slowed to a halt near the teacher's desk. Then he looked around the room at his classmates, as if speculating on his next activity. Noticing Gregory Daggett seated near

the back of the first grade, Raymond went directly to him. Raymond slid his knees into the seat immediately in front of Gregory, thus facing him. Raymond reached out and took a songbook from the top of Gregory's desk. *The two boys were extremely friendly in a very informal way which indicated mutual understanding. Gregory seemed to enjoy Raymond's presence.* (Barker and Wright 1951, pp. 71–72)

The preceding analysis took place over the course of a single minute, and the researcher involved took written notes as well as brief recorded statements. The italicized portion of the text appears that way in the original document and represents the observer's analysis. The viewer makes note of spatial minutiae, like how Raymond swung his arms and "jogged" to his teacher's desk. Alternatively, the viewer makes interesting observations about how the boy looked around the room, and even how the boy sat in the seat in front of his friend. All of these tiny details add up to describe the behavior of child locating a familiar friend, and informally engaging with them within the environment of the classroom. These details, in turn, show how the boy's actions are interrelated with the surroundings: the position of his teacher at the front of the room, the seats that allow him to slide his knees into the chair backwards thus facing his friend, or even how the boy swung his arms in the air.

Barker's contribution to our understanding of ecological methodologies is notable for many reasons, but perhaps most importantly, Barker's approach emphasizes an attention to the everyday occurrences happening in this young boy's life. His research is not about how the child coped with a loss in the family or moving to a new city. Nor does it revolve around discursive artifacts. Rather, Barker's study sought to examine the banal discursive and nondiscursive features of a situation, or the human and nonhuman ecology in an everyday experience, and from those details, put forth a plausible psychological explanation of observed behaviors.

The Ecology of Ecological Methodology: Garfinkel's Ethnomethodology

Barker's preoccupation with the everyday experiences of a child show how the quotidian is caught up with an ecological methodology. This connection is, all things being equal, a logical extension of ecological thinking. Ecological methods function on a logic of inclusion: to perform a study "ecologically" means to include factors in the study or analysis that one otherwise would have ignored, purposively disregarded, or downplayed. From the start, ecological methodology's scope of analysis was a forever-widening web, expanding to include new aspects of the ecology under investigation.

This was a central concern of Harold Garfinkel, whose 1967 *Studies in Ethnomethodology* complicated sociological methodology and raised questions pertinent to the development of ecological methodologies. As Garfinkel noted, sociology in the late 1960s had developed a focus on the familiar and common experiences of everyday life. However, despite a focus on everyday experiences, this "immense literature contains little data and few methods with which the essential features of socially recognized 'familiar scenes' may be detected and related to dimensions of social organization" (Garfinkel 1967, p. 36). To alleviate this problem, Garfinkel proposed what he called ethnomethodology (EM), an ethnographic approach to research that examines the everyday methodologies people utilize to navigate common situations.

For instance, Michael Mair et al. (2016) utilize EM in their study of Bayesian statisticians. They largely ignore the details of Bayesian statistics, and instead examine how these statisticians utilize "qualitative" frames of reference despite their intense focus on quantitative data. Asked to develop algorithms to analyze a database of education levels in different countries, these statisticians had to negotiate the variance in education systems by navigating different educational philosophies and practices. By examining the everyday practices of these statisticians' "data-crunching" for a client, the researchers argue that the "statisticians themselves were involved in the interpretative task of unraveling the interweaving of cultural and technical reasoning in the production of the artifacts they were working with" (Mair et al. 2016, p. 65). Utilizing EM, these scholars opened a space to think about how everyday practices can make the division between "quantitative" and "qualitative" disciplines less visible.

Garfinkel continued to develop EM and published several revisions that diverge significantly from the original model he developed in *Studies in Ethnomethodology*. Ilkka Arminen (2008), for instance, describes the cut between his early work and late work as the difference between a "scientific" and "radical" EM. In the former and earlier conception of EM, Garfinkel attempted to circumvent the lack of a methodology by insisting that analyses focus on the indispensably mundane, or the way in which members of a social group produce that group through recognizable methodologies. In later "radical" conception of EM, he suggests that even circumvention requires a methodology. Some of his early protégés, like Melvin Pollner, struggled with Garfinkel's break from his early formulation of EM, and lamented the dissolution of EM as a coherent methodological approach (Pollner 2012, p. 7).

Despite claims of ethnomethodology's end as a cohesive program, some have suggested that studies that use EM are broadly drawing on a tradition derived from Edmund Husserl's phenomenological approach and Alfred Schütz' concept of the "life world" (*lebenswelt*) (Cheng 2012, p. 584). Sometimes considered to be abstract and highly speculative, the phenomenological tradition can be conceived of as a coherent methodology for analysis. Liu Yu Cheng agrees, and provides a persuasive rationale: if EM "is regarded as a method, there is a boundary set, implicitly or explicitly, for the study of *ethnomethod,* limited to 'the rational properties' of practical actions" (Cheng 2012, p. 585). Or more simply stated, the highly phenomenological object of EM is "everything that happens and that is experienced around us" (Cheng 2012, p. 585). Since what happens is reportable, it must therefore have some rational properties. As such, these properties are made manifest in meaningful action, a "certain activity performed by any other social member for some reason to achieve some purpose" (Cheng 2012, p. 585). Cheng's analysis is a convincing argument for a baseline cohesiveness across the field of EM. However, Garfinkel's postmodern reflections on EM are justified: no methodology should be considered *the* methodology and theoretical rigidity can produce oversights in reasoning (recall the cultural evolutionists). However, systematizing the basic tenets of phenomenological analysis can be another way to approach an ecological methodology, as Garfinkel demonstrates. The object of phenomenology's gaze is nothing other than an ecology.

Garfinkel's EM has led us down theoretical paths that cut to the core of what it means to do research on human subjects. Unlike the botanists, biologists, and geneticists that employ their methods to systematically examine the natural world, the study of human behavior, interaction, and experience requires a different set of methodological tools that can account for the unruly and unpredictable structures of everyday human life. The thinkers, methods, and approaches to ecological thinking we have discussed thus far show how varied ecological approaches can be: the proto-ecologists demonstrated how influential a hierarchical state of mind can be in conceptualizing an ecology, just as they showed how an ecological state of mind is viable for a large and varied set of disciplinary pursuits. Alternatively, the ecological methods developed by sociologists, anthropologists, and psychologists in the twentieth century produced a multifaceted understanding of human experience.

Much more could be said about how far and wide the idea of ecology has traveled. Indeed, whole volumes could be composed of the various histories of ecological thinking over a wide selection of disciplines and sub-disciplines. We have not attempted to produce this kind of in-depth analysis. Instead, the preceding can be thought of as an ecology of the idea of ecology, and the methods those ideas give birth to. Since an ecology is not required to be ordered hierarchically, we suggest that no one movement, thinker, or idea in this ecology of ecologies is more or less influential than any other. Fractal-like, these ideas are caught up in their own historically situated ecologies, which affect their relationship to themselves and one another. This kind of ecological relationship can be drawn nearly *ad infini-tum* unless we impose some kind of domain or limitations on our practice of ecological thinking. In constructing this methodology, however, we should remember the risk of environmental determinism, coloring our observations with our own biases and privileging some features over others without a rationale. We argue that a middle ground is possible.

THE ORIGINS AND BOUNDARIES OF ECOLOGICAL VALIDITY

We propose a third way of utilizing ecology, one that is neither so broad and abstract that it includes all things at all times nor reductive to the point of self-defeat. This middle ground can be operationalized in the criterion of ecological validity. Researchers sometimes mistake ecological validity for a kind of external validity. However, while external validity asks whether results can be generalized across experimental settings, ecological validity asks whether they apply to existing ecologies (Breau and Brook 2007). In addition, while external validity and ecological validity are related, they are independent. David Breau and Brian Brook (2007) define ecological validity as "the degree of similarity between the conditions of a simulation experiment and the real-world phenomenon that the experiment is designed to model" (p. 78). Similarly, Pat Gehrke defines ecological validity as "the degree of correspondence between the research conditions and the phenomenon being studied as it occurs naturally or outside of the research setting" (Gehrke forthcoming).

Ecological validity's origins can be traced back to the works of Brunswik and Lewin in the 1940s. Brunswik is considered the first to utilize the term "ecological validity," but his version describes a narrower conception of ecological validity than what is utilized today and what we mean by the term. According to Mandeep Dhami et al. (2004), ecological (or predictive) validity for Brunswik was a measure of the correlation between a cue and the distal variable in a situation. For example, "when considering the cues that a judge may use to predict absconding on bail, the defendant's gender may be highly related to the type of offense with which he or she is charged" (Dhami et al. 2004, p. 961). Or, if the defendant is being charged with committing a sexual offense and the defendant is male, since men are more likely to commit this type of offense the correlation between gender and whether or not the defendant will abscond on bail demonstrates strong ecological validity. Some, like Kenneth Hammond, have noted that what has become modern ecological validity is actually a conflation of Brunswik's concept of ecological validity and another related concept called "representative design" (Hammond 1978, p. 4). The conflation is perhaps confounding, but regardless of this denotative shift, when we say ecological validity we mean a concept more in line with the definition given by Breau and Brook, or the work associated with Lewin, who was a contemporary of Brunswik.

In 1943, Lewin responded to Brunswik's conception of a person's "life-space," or the "field" that researchers have in mind when they refer to motivations, moods, goals, anxieties, ideals, or needs of a subject (Lewin 1943, p. 306).[2] Where Brunswik and Lewin depart is in how they conceptualize a life-space's limits: Brunswik believed that it should not be confused with the environment or actually achieved results in the environment, while Lewin believed that a life-space can be captured by a psychological field at a given time. In other words, Lewin sees what the subject is perceiving in their visual field and their motivations as constituting a life-space. However, he does "not consider as part of the psychological field at a given time those sections of the physical social world which do not affect the life space of the person at that time" (Lewin 1943, p. 307). He explains: "The food that lies behind the doors at the end of a maze so that neither smell nor sight can reach it is not a part of the life space of the animal. In case the individual knows that food lies there this *knowledge*, of course, has to be represented in his life space, because this knowledge affects behavior. It is also necessary to take into account the subjective probability with which the individual views the present or future state of affairs because the degree of certainty of expectation also influences his behavior" (Lewin 1943, p. 307). We can glean two important ideas from this example. First, Lewin establishes a boundary suggesting that what defines the scope of a given ecology (boundary included) as what is constrained by the sensorial field at a given time. Second, Lewin seems to intuitively understand that a life-space is always already a term loaded with ecological meaning. Both human (thoughts, wishes, dreams, etc. that are relevant for an observed behavior at a given time) and nonhuman (sensorial features of a scene at a given time) elements are interwoven, producing an ecology.

What do Lewin's boundaries mean for ecological validity? If a study demonstrates strong ecological validity, the data produced will account for the common-sense intuition that if we wish to examine human behavior, we should replicate the conditions under which behaviors actually occur. By "actually occur," we simply mean that research studies should replicate the conditions under which a subject would behave in their living world. Lewin provides two key insights: first, a living world is capable of being systematized, and, second, a person's living world includes material and non-material, human and nonhuman components insofar as they are purposeful at a given time. The first insight is important because it affords us the capacity to suggest there are concrete limits to what constitutes an ecology at a given time, and because ecological validity asks that research replicate the ecological conditions under which a behavior occurs, realist criticisms that ecological validity is too abstract to be useful can be discarded. The second insight is important because it allows us to argue that a vast range of material and non-material forces influence behavior in a given ecology at a given time, and because ecological validity asks that research replicate the ecological conditions under which behavior occurs, idealist criticisms that ecological validity does not conceptually account for the full richness of an ecology can also be discarded. In either case, Lewin's contribution provides us with a way to implement ecological thinking and the criterion of ecological validity as practical guides for public engagement research.

For many years, researchers across disciplines have been calling for greater attention to ecological validity. Recall the work of Barker, who was concerned with the same problems as Brunswik and Lewin, mainly that lab conditions did not reflect the actually occurring world. Barker's study had strong ecological validity because he responded to this problem by performing psychological fieldwork. Given Barker's preoccupation with childhood development, this was probably the only way in which he could maintain adequate ecological validity in a study of a child's everyday life. However, Gibson's *The Ecological Approach to Visual Perceptions* articulates a commitment to ecological validity when performing experiments about a very different kind of ecology, visual perception.

Gibson's (1979) demonstration of strong ecological validity was a rejection of the standard stimulus-response methods traditional psychological studies of optics relied on. At the time, the equipment being used in labs to study optics focused on narrow visual processes, like aperture vision. This equipment had, in turn, made it seem like optics researchers could *only* study these narrow processes. Gibson disagreed, and ventured to study natural vision, the kind of vision that we utilize in everyday situations (Gibson 1979, p. 3). Unlike Barker, whose research had strong ecological validity by doing psychological fieldwork, Gibson's interest in perception affords him the possibility of strong ecological validity in the lab. As Gibson exclaims, "The laboratory *must* be life like!" (Gibson 1979, p. 3). Together, Gibson and Barker demonstrate how strong ecological validity is achievable in both lab and field settings.

Gibson's work influenced Neisser, whose publication of *Cognitive Psychology* in 1967 fueled his meteoric rise as the "father" of cognitive psychology. Neisser was actually a good friend with James Gibson and his wife Eleanor Gibson, whose body of scholarship is as influential and insightful as James's, if not more so.[3] Neisser's relationship with these two scholars prompted him to write *Cognition and Reality* in 1976. It did not receive that same warm reception as *Cognitive Psychology*, in large part because Neisser incorporated an ecological dimension into his cognitive paradigm. Instead of focusing on vision or child development, Neisser devoted his attention to memory. He admitted in a 1997 interview that Gibson's insistence on strong ecological validity guided much of this new research, where he did various "ecological studies of memory" (Szokolszky 2013, p. 196). When asked what Neisser saw as the future of ecological methodology in psychology, he remarked the "environment is just as real as the brain and ecological psychologists seem to be the only people who work that side of the street consistently and make discoveries there" (Szokolszky 2013, p. 199).

Since at least the 1970s, ecological validity has also played an important role in jury studies (Bornstein 1999, p. 75). These studies hold a substantial amount of sway in the legal system, and psychological research has been used in such rulings as capital punishment, the effects of pretrial publicity on cases, and rules concerning jury decisions (Bornstein 1999, p. 76). The extent to which the findings of jury studies are applicable to actual settings is, given this importance, crucial. After all, a person's property, freedom, or even life might hang in the balance. Indeed, the Supreme Court has chosen to discount research on the basis of a perceived lack of validity (Bornstein 1999, p. 76). At the intersection of law and psychology, we see how important a measure of ecological validity might be: if studies are to be used to make claims about reality, those studies should attempt, as much as possible, to replicate the ecology in which that reality unfolds.

Ecological Validity and Public Engagement with Science

We have seen how ecological validity began as a concept in psychology and how it was used by Brunswik, Lewin, Gibson, Neisser, and scholars interested in jury studies. However, our interest in this book is in how ecological thinking can improve public engagement with science. Hence, the rest of this book articulates how ecological validity (and more broadly, ecological thinking) can change how scholars in public engagement with science conduct research and measure the validity of their results. As Richard Jones (2007) argues, public engagement with science is caught between competing objectives. On the one hand, the cynical perspective on engagement is "to defuse potential public opposition" and "obtain some fig leaf of public consent to a decision that has already been made" (Jones 2007, p. 262). Nick Pidgeon and Tee Rogers-Hayden (2007) echo this criticism, claiming much of public engagement with science only occurs late in research and development, "after a controversial ethical risk of question has arisen" (p. 194). On the other hand, the idealist position is "to have a real dialogue with the public about a future that remains genuinely open" (Jones 2007, p. 262). This "real dialogue," in practice, reflects what Darrin Hicks (2002) and Darrin Durant (2011) identify as deliberative democracy theories informed by the work of Jürgen Habermas and John Rawls.[4] Public engagement guided by such philosophies of democracy tends to impose idealized or at least highly guided models of communication and discussion, designed to "improve" the quality of discourse or more effectively produce agreement.

Neither the cynical nor the idealist perspectives emphasize understanding existing publics, except as such understanding mediates between limited-scope events and broader cynical or idealist purposes. In either case, publics are prefigured as faulty; they must either be led to the correct conclusion (cynical) or be taught the right way to deliberate (idealist). Torn between these competing motivations for public engagement work, we lose sight of the role public engagement can play in understanding the reality of publics and how they come to opinion and action. Researchers in the field of public engagement with science have attempted to represent publics via standardized methods at the expense of methods that promise higher ecological validity, largely missing how and why actual publics engage with issues. Additionally, while public engagement scholars have long expressed concern with scientific experts' efforts to control public conversations about science, only rarely do they reflect that concern back on the public engagement experts' own tendency to dictate the form and content of public engagement events.

Daniel Kleinman (2000) has noted that approaches to studying public engagement with science vary in the degree of scientific expert control of engagement, ranging from scenarios in which "scientists defined the agenda, marginalized lay input, and excluded non-technical matters" to grassroots social movements that inserted themselves into discussions of science and science policy. While this spectrum has become foundational to public engagement with science, the current crisis in public engagement emerged from a corollary problem of control by a growing body of public engagement experts who artificially govern the conditions of engagement. Our solution to this crisis is a method of public engagement that prioritizes ecological validity and is informed by research in vernacular rhetoric, quasi-ethnography, and the role of social context in the public understanding of science. As we have explored throughout this chapter, these goals are not new, but their application in the field of public engagement with science is novel, as is our particular method of articulating this methodology through use of scholarship in citizenship theory, deliberative democracy, and rhetorical studies. We call this methodology *organic public engagement*.

In order to provide criteria for ecological validity in public engagement with science studies, we look to research in jury studies. Kleinman (2000) has noted the relevance of jury research to public engagement and particularly public understanding of science. Both fields deal with learning or assimilating new and often technical or specialized information, interacting with others who have varying relations to the issue under debate, and often require collective decision-making or action. As Brian Bornstein (1999) has articulated, ecological validity in jury studies involves five dimensions of research design: setting, sample, communication medium, amount and type of deliberation, and perceived consequences of the deliberation (pp. 75–76). These standards are significant for public engagement studies, and when combined with work in comparative cognitive psychology, offer a path toward improving our methods. In the following sections, we explore these standards and provide examples of weak and strong ecological validity.

Setting

The first standard for ecological validity is setting, which provides the basis for sampling. Setting plays an important role because its alteration changes the outcomes of research. However, an ecologically valid setting includes not only the physical environment, but also the broader social environment in which a task or behavior occurs. Recall our discussion of Lewin's concept of the life-space: the boundary of a given ecology is defined both by those components that fall within the sensory field of the participant and the motivations, ideas, or purposes that guide the participant's action at the time of observation. In addition, as Michael Cole et al. (1997) argue, ecological validity requires not only studying people in actually occurring situations but also allowing problems and questions to emerge from within these situations. Citing a study on memory which asked passersby on a university campus "What day is today?" and then recorded the reaction time for response, they note how researchers did not "discover" individuals being asked what day of the week it was. Nor did they observe their responses when participants in the study were asked this question while confronting a researcher with a clearly visible stop-watch. A researcher cannot enter an ecology and introduce problems through means not endemic to that life-space without altering the ecology, therefore, reducing the study's validity. A related problem is demonstrated by Jean Lave's study on tailors' arithmetic in their daily work. He found that when arithmetic problems were posed to tailors as part of a study with instructions, their means of solving those problems differed from how they handled arithmetic in their daily work. Lave (1997) concludes that traditional experimental methods designed to produce commensurable and quantifiable objects of analysis may be incompatible with some of the everyday ways in which people solve problems.

Julie Williamson and Daniel Sundén's (2015) research on human-computer interaction is an example of a study where setting reflects strong ecological validity. Their "deep cover" methodology, which they describe as turning "to the wild," has four key components: it blurs the lines between experimental and real world settings, the experimenter has no intervention of visible presence after the initial intervention, analysis is based on multiple sources of observable data only, and there is no explicit consent gathered from participants at any point during the study (Williamson and Sundén 2015, p. 543). These four components comprise a methodology that seeks to access the conditions under which people actually use computers.

K.C. O'Doherty and A. Hawkins's (2010) research on policy development for human tissue biobanking offers a more complicated relationship to ecological validity. Their methodology incorporates a complex deliberative model that in many ways demonstrates an awareness of the problems with weak ecological validity. However, one central tool used during their research involved the development of a "deliberation workbook," which (along with a basic description of the topics involved in the study), included "vignettes, explanations of relevant terminology, examples of recommendations from the previous (unstructured) public engagement on biobanking... and recognized pros and cons" (O'Doherty and Hawkins 2010, p. 202). Introducing both examples of recommendations from previous engagements and recognized pros and cons can pose problems and present sources of information not endemic to the ecology, weakening the study's ecological validity if its objective is to represent opinions or opinion formation as it exists outside the research setting. As Gehrke put it, "If one purports to study participants' knowledge and opinions of a scientific theory, then interference with their usual channels of information and methods of collectively or individually coming to opinion will dramatically weaken the ecological validity of the research" (Gehrke forthcoming).

Sample

The second criterion and a major implication of research on setting is how we sample publics. As Cole, Hood, and McDermott summarized, ecologically valid research changes sampling in two ways: it takes "situations, or tasks, rather than people" as the "basic units of analysis" and manages sampling and representativeness by insisting upon a diversity of sites and situations (Cole et al. 1997, p. 50). This method of sampling requires that researchers identify the task or behavior being studied, locate the diversity of environments in which that task or behavior occurs, and then sample widely from those environments. As Kleinman (2000) has noted, common public engagement models such as citizen juries tend to approach samples as aggregate demographic cross sections of a broad population, but such sampling alters the studied subjects by displacing them from their constitutive settings. Each subject is expected to represent the demographic and psychographic domain from which they were plucked, under the erroneous assumption that displacement and insertion into a new setting will not alter attitudes or actions. Rather than sampling individuals, an ecologically valid mode of research samples environments, ecologies, and situations in which the object of study occurs or might occur independent of researcher intervention.

A good example of stronger ecological validity in sampling can be seen in research of Troy Hartley and Robert Robertson (2006). Their study on the Northeast Consortium evaluates how this organization encouraged partnerships between scientists, individuals engaged in commercial fishing, and industry members who participated in a variety of fisheries and cooperative research projects. Their selected survey data attempted to capture all three of these group's opinions on the benefits of the Northeast Consortium, which ranged widely in education level, experience, and knowledge base, but nonetheless all participated in different aspects of the fishing industry (Hartley and Robertson 2006, p. 166).

Devidas Menon and Tania Stafinski's (2008) research on health technology assessment, however, demonstrates weaker ecological validity. Their study selected 1600 randomly sampled residents of a specific region in Alberta, Canada and asked them to deliberate on the value of new healthcare technologies. These technologies ranged from treatments of obesity to arthritis, hearing problems, colon cancer, depression, and others. Despite the wide variety of technologies presented to these citizen juries, this study only selected participants that are "broadly representative of their community" (Menon and Stafinski 2008, p. 284). Each of these citizens is a member of multiple ecologies, multiple "publics," and they will communicate and reason differently in different contexts. It seems reasonable to suggest that the Deaf community might see and understand technologies that affect hearing loss very differently than a random sampling of citizens, and a Deaf citizen is likely to articulate their views differently to strangers, especially a group of hearing citizens in contrast to the Deaf community. Regardless of the deliberative nature of this study, the results do not capture how publics with different stakes in the technologies presented would assess their efficacy and value.

Communication Medium and Deliberation

While potentially included within setting, communication medium and characteristics of deliberation are of special concern for both jury studies and public engagement research. Jury studies have found significant differences between responses to trial transcripts, videos, and live reenactments or simulations. Likewise, altering the type or duration of deliberation changes the outcome. As a result, experimental designs that specify time constraints or means of deliberation that do not mirror the practices of actual juries have weaker ecological validity. One may, however, treat deliberation as an independent variable and study how alterations in deliberation change outcomes, but such research studies counter-factual situations, usually for the purpose of proposing changes in jury instructions or courtroom procedures. If, instead, one is studying the practices of juries as they exist, jury research documents the importance of mirroring every facet of jurors' experiences as closely as possible or, ideally, gaining access to actual juries in the process of trying cases.

Altering modes of communication and deliberation alters participants' systems of valuing, identifying, and relating, which is precisely why deliberative democracy idealists force such alterations in existing publics and seek to remake publics in the image of their political philosophies. Steven Epstein's research on AIDS activists demonstrates how medium, deliberation, and setting together alter the representativeness of public engagement with science. The contribution of nonexpert publics, according to Epstein (2000), involves how they think, communicate, and place value differently from scientific communities and thus potentially transform or complement the practices of scientists. Epstein warns that "insofar as activists start thinking like scientists and not like patients, the grounding for their unique contributions … may be in jeopardy of erosion" (Epstein 2000, p. 25).

James Fairhead et al.'s (2006) ethnographic study of a vaccine trial performed in Gambia has strong ecological validity in its selection of medium and its treatment of deliberation. While they found that fieldworkers often interpreted local citizen's refusal to participate as a sign of ignorance, tradition, religion, wealth, or ethnicity, citizen's actual reasons for abstaining from the study were often varied, complex, and tied to a deliberative process (Fairhead et al. 2006, p. 115). As Fairhead, Leach, and Small note, much contemporary discussion revolves around the implications of "transnationally constituted" science on local populations and how they struggle with this intervention. While they admit that such struggles do occur, they caution that these are not the *only* kinds of interactions that occur with healthcare practices. Their study demonstrates strong ecological validity by considering how subjects deliberated in a vaccine study, and how that deliberation affected their willingness to coalesce with the trial. In doing so, they more closely reflect how healthcare decisions are made in this particular ecology.

Perceived Consequences

Finally, the fifth criterion for ecological validity is the perceived consequence of deliberation. Breau and Brook (2007) found that when juries were told they were participating in simulations, they were harsher in both determining guilt and recommending punishments than when they believed they were trying actual cases (p. 78). Similar results have been found in numerous contexts, such as work by Robert Ladouceur et al. (1991) on ecological validity and gambling, where they found that gamblers bet more money in the laboratory than in their "natural" settings, as the outcome of their gambling is materially different. The lesson for public engagement is clear: if participants perceive different implications for their simulated deliberations than in organic situations, then their behaviors and their decisions may differ. While public engagement scholars usually understand consequences as influence on policy or action, they may also extend to membership, affiliation, or social belonging, which members risk when they speak and act in their daily lives. Thus, we ought to match perceived implications in our studies to perceived stakes in real publics, including the stakes for social relations and community membership.

Readers familiar with common methods used in public engagement with science are no doubt already seeing the transformative implications of ecological thinking and ecological validity. With this brief history as background, and now equipped with a sufficiently robust model for assessing ecological validity, we turn in Chap. 3 toward applying that model to the most common practices of public engagement with science. In doing so, we offer examples of approaches that offer both weaker and stronger ecological validity in their research design.

NOTES

1. For an example of this research, see Beach, Steven R.H., Gene H. Brody, Alexandre A. Todorov, Tracy D. Gunter, and Robert A. Philibert, "Methylation at SLC6A4 is linked to Family History of Child Abuse: An Examination of the Iowa Adoptee Sample," *American Journal of Medical Genetics Part B: Neuropsychiatric Genetics* 153, no. 2 (2010), 710.
2. Lewin, Kurt, "Defining the 'Field at a Given Time'," *Psychological Review* 50, no. 3 (1943), 306 (292–310). It's not entirely clear, but it seems that Lewin is referencing "life world" (recall Garfinkel was influenced by Schütz use of *lebenswelt*), so we should probably clarify that this term originates with

Edmund Husserl. Husserl first used the word in his *Crisis of the European Sciences* published partially in 1936, and partially later in 1954. Husserl's use of the term *lebenswelt* is used to describe how European science had taken on a perspective grounded solely in purely "objective" or mathematical law. As Husserl defines it, the *lebenswelt* is both pre-behavioral and pre-theoretical and therefore resistant to mathematical representation—a kernel of reality before action and before conceptualization. Brunswik tentatively agrees with this presupposition, but Lewin thinks that this zone is methodologically "closeable."

3. See E.J. Gibson's 1960 "The 'visual cliff'," her most famous experiment where she demonstrated that many (but not all) animals have an innate ability to perceive depth.

4. See Darrin Hicks, "The Promise(s) of Deliberative Democracy," *Rhetoric & Public Affairs* 5, no. 2 (2002), 223–260, and Darrin Durant, "Models of Democracy in Social Studies of Science," *Social Studies of Science* 41 (2011), 691–714.

References

Arminen, Ilkka. 2008. Scientific and 'Radical' Ethnomethodology. *Philosophy of the Social Sciences* 38 (2): 167–191.

Barker, Roger G. 1968. *Ecological Psychology: Concepts and Methods for Studying the Environment of Human Behavior*. Stanford: Stanford University Press.

Barker, Roger G., and Herbert F. Wright. 1951. *One Boy's Day: A Specimen Record of Behavior*. New York: Harper and Brothers.

Bornstein, Brian H. 1999. The Ecological Validity of Jury Simulations: Is the Jury Still Out? *Law and Human Behavior* 23 (1): 75–91.

Breau, David L., and Brian Brook. 2007. 'Mock' Mock Juries: A Field Experiment on the Ecological Validity of Jury Simulations. *Law and Psychology Review* 31: 77–92.

Cheng, Liu Yu. 2012. Ethnomethodology Reconsidered: The Practical Logic of Social Systems Theory. *Current Sociology* 60 (5): 581–598.

Cole, Michael, Lois Hood, and Raymond McDermott. 1997. Concepts of Ecological Validity: Their Differing Implications for Comparative Cognitive Research. In *Mind, Culture, and Activity: Seminal Papers from the Laboratory of Comparative Human Cognition*, ed. Michael Cole, Yrjö Engström, and Olga Vasquez, 49–56. Cambridge: Cambridge University Press.

Corsi, Pietro. 1988. *The Age of Lamarck Evolutionary Theories in France 1790–1830*. Berkeley: University of California Press.

DeLisi, Matt, and Michael G. Vaughn. 2015. The Vindication of Lamarck? Epigenetics at the Intersection of Law and Mental Health. *Behavioral Sciences and the Law* 33 (5): 607–628. https://doi.org/10.1002/bsl.2206.

Dhami, Mandeep K., Relph Herwig, and Ulrich Hoffrage. 2004. The Role of Representative Design in an Ecological Approach to Cognition. *Psychological Bulletin* 130 (6): 959–988.

Dove, Michael R. 2015. Linnaeus' Study of Swedish Swidden Cultivation: Pioneering Ethnographic Work on the 'Economy of Nature'. *AMBIO* 44 (3): 239–248. https://doi.org/10.1007/s13280-014-0543-6.

Durant, Darrin. 2011. Models of Democracy in Social Studies of Science. *Social Studies of Science* 41 (5): 691–714.

Epstein, Steven. 2000. Democracy, Expertise, and AIDS Treatment Activism. In *Science, Technology, and Democracy*, ed. D. Daniel Kleinman, 15–32. Albany: Sate University of New York Press.

Fairhead, James, Melissa Leach, and Mary Small. 2006. Public Engagement with Science? Local Understandings of a Vaccine Trial in the Gambia. *Journal of Biosocial Science* 38 (1): 103–116.

Feinberg, Andrew P., and Bert Vogelstein. 1983. Hypomethylation Distinguishes Genes of Some Human Cancers from Their Normal Counterparts. *Nature* 301 (6): 89–92.

Foucault, Michel. 1970. *The Order of Things*. New York: Vintage Books.

Garfinkel, Harold. 1967. *Studies in Ethnomethodology*. Englewood Cliffs: Prentice-Hall.

Gehrke, Pat J. forthcoming. Ecological Validity. In *The SAGE Encyclopedia of Educational Research, Measurement, and Evaluation*, ed. Bruce Frey. Thousand Oaks: SAGE.

Gibson, James J. 1979. *The Ecological Approach to Visual Perception*. Boston: Houghton Mifflin Company.

Haeckel, Ernst. 1866. *Generelle Morphologie*. Berlin: Georg Reimer. Quoted in Robert J. Richards. 2008. *The Tragic Sense of Life: Ernst Haeckel and the Struggle over Evolutionary Thought*, 144. Chicago: University of Chicago Press.

Hammond, Kenneth R. 1978. Psychology's Scientific Revolution: Is It in Danger? *Center for Research on Judgment and Policy* 211: 1–45.

Hartley, Troy W., and Robert A. Robertson. 2006. Stakeholder Engagement, Cooperative Fisheries Research and Democratic Science: The Case of the Northeast Consortium. *Human Ecology Review* 13 (2): 161–171.

Hicks, Darrin. 2002. The Promise(s) of Deliberative Democracy. *Rhetoric & Public Affairs* 5 (2): 223–260.

Hutton, James. 1788. *Theory of the Earth*. Edinburgh: The Royal Society of Edinburgh.

Jones, Richard. 2007. What Have We Learned from Public Engagement? *Nature Nanotechnology* 2 (5): 262–263.

Judkins, Gabriel, Marissa Smith, and Eric Keys. 2008. Determinism Within Human-Environment Research and the Rediscovery of Environment Causation. *The Geographical Journal* 174 (1): 17–29.

Kleinman, Daniel. 2000. Democratization of Science and Technology. In *Science, Technology, and Democracy*, ed. Daniel Kleinman, 139–165. Albany: State University of New York Press.

Koerner, Lisbet. 1999. *Linnaeus: Nature and Nation*. Cambridge: Harvard University Press.

Ladouceur, Robert, Anne Gaboury, Annie Bujold, Nadine Lachance, and Sarah Tremblay. 1991. Ecological Validity of Laboratory Studies of Videopoker Gaming. *Journal of Gambling Studies* 7 (2): 109–116.

Lave, Jean. 1997. What's Special About Experiments as Contexts for Thinking. In *Mind, Culture, and Activity: Seminal Papers from the Laboratory of Comparative Human Cognition*, ed. Michael Cole, Yrjö Engström, and Olga Vasquez, 57–69. Cambridge: Cambridge University Press.

Lewin, Kurt. 1943. Defining the 'Field at a Given Time'. *Psychological Review* 50 (3): 292–310.

MacDonald, Dennis W. 2011. Beyond the Group: The Implications of Roderick D. McKenzie's Human Ecology for Reconceptualizing Society and the Social. *Nature and Culture* 6 (3): 268–284.

Mair, Michael, Christian Greiffenhagen, and W.W. Sharrock. 2016. Statistical Practice: Putting Society on Display. *Theory, Culture & Society* 33 (3): 51–77.

McKenzie, Roderick D. 1924. The Ecological Approach to the Study of the Human Community. *American Journal of Sociology* 30 (3): 287–301.

Menon, Devidas, and Tania Stafinski. 2008. Engaging the Public in Priority-Setting for Health Technology Assessment: Findings from a Citizen's Jury. *Health Expectations* 11 (3): 282–293.

O'Doherty, Kieran, and Alice Hawkins. 2010. Structuring Public Engagement for Effective Input in Policy Development on Human Tissue Biobanking. *Public Health Genomics* 13 (4): 197–206.

Orlove, Benjamin S. 1980. Ecological Anthropology. *Annual Review of Anthropology* 9 (23): 235–273.

Park, Robert E., Ernest W. Burgess, and Roderick D. McKenzie. 1925. *The City*. Chicago: University of Chicago Press.

Pidgeon, Nick, and Tee Rogers-Hayden. 2007. Opening Up Nanotechnology Dialogue with the Publics: Risk Communication of 'Upstream Engagement'? *Health, Risk & Society* 9 (2): 191–210.

Pollner, Melvin. 2012. The End(s) of Ethnomethodology. *American Sociology* 43 (1): 7–20.

Reynolds, Andrew. 2008. Ernst Haeckel and the Theory of the Cell State: Remarks on the History of a Bio-Political Metaphor. *History of Science* 46 (2): 123–152.

Richards, Robert J. 2008. *The Tragic Sense of Life: Ernst Haeckel and the Struggle Over Evolutionary Thought*. Chicago: Chicago University Press.

Steward, Julian H. 1972. *Theory of Culture Change: The Methodology of Multilinear Evolution*. Champaign: University of Illinois Press.

Szokolszky, Agnes. 2013. Interview with Ulric Neisser. *Ecological Psychology* 25 (2): 182–199.

Williamson, Julie R., and Daniel Sundén. 2015. Deep Cover HCI: A Case for Covert Research in HCI. In *CHI EA 15: 33rd Annual ACM Conference Extended Abstracts on Human Factors in Computing Systems*, April 18–23, 543–554.

Wulf, Andrea. 2015. *The Invention of Nature: Alexander Von Humboldt's New World*. New York: Alfred A. Knopf.

CHAPTER 3

Engaging Public Ecologies

Having laid out the salient features of ecological validity in Chap. 2, this chapter provides readers an overview of research methods often classified as forms of public engagement. We begin with survey and media studies, where communication is analyzed from one direction. Survey studies are what Gene Rowe and Lynn Frewer (2005) call "public consultation," which is characterized by information being transferred from a public to an interested organization or agency. Media studies are what Rowe and Frewer (2005) characterize as "public communication," because they analyze how information is transferred from organizations to a public. We call these engagements "unengaged" because the public and the regulatory or governmental agency are not in a participatory dialogue.

After presenting the strengths and weaknesses of survey and media studies with regards to ecological validity, we shift to more ambiguous sites of engagement, which include oral, curatorial, and digital spaces. Unlike survey studies and media studies, these events demonstrate more potential engagement. Despite this potential, little research has been done on how these sites contribute to publics' understanding of a diverse range of topics. Therefore, in this section we both review the ecological validity of current literature on these spaces and provide suggestions for how future research can take account of ecological concerns.

Finally, we turn to what Rowe and Frewer (2005) call "public participation." Unlike public communication or consultation, public participation is a deliberative event that occurs between varying sizes of public

© The Author(s) 2018
A.S. Lerner, P.J. Gehrke, *Organic Public Engagement*,
https://doi.org/10.1007/978-3-319-64397-7_3

participants and scientists, organizations, agencies, or interested parties. Most models for this deliberative process assume that the negotiation between public and sponsor changes the opinions of both groups. We assess the ecological validity of four kinds of participatory engagement events: consensus conferences, citizen panels, citizen juries, and deliberative polling. For each, we examine relevant research studies and explain their commitment (or lack thereof) to ecological validity.

Our argument in this chapter is that very little research currently done in the field of public engagement possesses strong ecological validity. Survey and media studies present some opportunities for the consideration of ecological validity, however, by design, surveys fail to capture how publics encounter, utilize, synthesize, or otherwise engage with information. Alternatively, media studies often infer characteristics of the public, when what they actually capture is the ecology of journalistic practices. Oral, curatorial, and digital engagements are more suited for engagement research, yet current research on these spaces either focuses on a narrow space or time frame, or is primarily concerned with the number of participants. Neither of these options lends itself to strong ecological validity. Deliberative engagements are by far the most popular, novel, and widely researched sites for understanding public engagement; however, these events usually utilize artificially constructed spaces of engagement and therefore studies that infer public opinions or beliefs from these events have weak ecological validity. In Chap. 4 we further explain the origins and reasoning behind the ecological failure of deliberative engagement.

Importantly, a study with weak ecological validity does not imply logical deficiency, ethical inferiority, or poor design. On the contrary, much of the work being done by public engagement scholars is novel, valuable, thorough, and important for organizations who wish to better understand how publics think about topics in contemporary society. Instead, a majority of these studies could be more robust, useful, and persuasive if they better attended to the ecological validity of their research design. In the spirit of ecological thinking, we believe that ecological validity is an additive principle to public engagement methods: when ecological validity is considered, the data gathered during a study will capture a more robust representation of reality, which in turn can lend that data a persuasive weight it would otherwise lack.

Additionally, the kinds of public engagement methods that we have in mind cross several different disciplines. Public health, geoengineering, medical technologies, and child services all deploy public engagement methods, and we cite widely from a variety of literatures to demonstrate how ecological validity can be applied. However, we focus primarily on studies that fall under the banner of science communication, health communication, or science and technology studies. While these fields have distinct scholarships, they are connected by a shared concern with how rarefied information is processed, understood, resisted, and utilized by publics.

UNENGAGED ENGAGEMENTS

One of the most coherent sites for better understanding public engagement is the journal *Public Understanding of Science* (*PUS*). The exigency for *PUS* emerged in the UK during the mid-1980s, where the scientific establishment anxiously responded to "a legitimation vacuum which threatened the well-being and social standing of science" (Wynne 1992, p. 38). By the mid-1990s, a group of scholars had begun to coordinate their efforts to better understand this phenomenon. As Brian Wynne (1992) suggests in the first issue of *PUS*, this vacuum emerged from science distancing itself from the "ordinary public" (p. 38). This distancing was equated with a public lack of *understanding* of science. Therefore, scholars assumed that a perceived "opposition" to science was caused by a lack of understanding of science's goals, methods, and aspirations.

Over time, the emphasis for these researchers moved from understanding to engagement, which gave life to what we would now call "public engagement with science." Hence, public engagement with science began to develop as a specific application of public engagement methods. One long-lasting debate in public engagement with science is the question of the public's scientific literacy, or more specifically the connection between literacy and attitude. This debate finds traction in the very first issue of *PUS*, where scholars discuss the "deficit model" of scientific literacy, which assumes that public ignorance is positively correlated with a lack of public support for a variety of issues in science, technology, and health. Much of the support for this view came from survey studies.

Survey Studies

The first and simplest form of public engagement is the survey. Surveys build a snapshot of a public by aggregating individual responses, usually to determine public attitudes and correlations between those attitudes and/or demographics. They often purport to provide insight into both how opinions are distributed in the public and how varying attitudes, demographics, or psychographics might be related. Most of these instruments constrain respondents to set choices, such as multiple-choice or Likert-type scales, and when a respondent refuses the given schema for responses, researchers most commonly force responses into an existing category or discard the response altogether. For example, a survey may use a Likert scale asking for a response from 1 to 4, leaving no middle or neutral option, but respondents may write-in 2.5 or circle both 2 and 3, and survey researchers tend to either discard those responses or normalize them by rounding. Likewise, surveys often ask for respondents' sex or gender and give only two options (male or female). Researchers then often discard any attempts by respondents to articulate a different gender category.

Most survey studies are at their highest point of ecological validity when they study actions normally performed in conditions that mirror those in which respondents complete the survey: isolation, anonymity, privacy, aggregation, and forced limited choice (such as voting). Alternatively, survey studies can produce certain baselines about the individuals that populate a public, given they use a robust set of sampling methods that examine actual stakeholders. However, surveys are poor methods for representing how publics respond to new information, form opinions, or are persuaded as communities or associations as they actually exist. As we have argued in Chap. 2, studies that have examined the effects of setting on research outcomes are sufficient to warrant skepticism of most survey studies, particularly when it comes to questions of new information and risk messages.

One of the major assumptions built upon survey data is public engagement with science's development of a deficit model of scientific literacy. According to this model, "science is seen as a well-defined body of knowledge, and the public is judged according to how much of this knowledge it possesses" (Durant et al. 1992, p. 162). John Durant, Geoffrey Evans, and Geoffrey Thomas's (1992) article defending this model is an excellent example of how a survey can demonstrate weak ecological validity. Their study, which does not note the sample population or setting, utilized a survey given out in the UK that asked several quiz-like statements, which respondents marked either "true," "false," or "don't know."

While the responses shed light on some interesting gaps in public knowledge, what Durant, Evans, and Thomas's survey approach fails to capture is the ecology of *why* a given public might resist a quiz-like survey, or why they might willfully demonstrate ignorance. Brian Wynne's (1992) quasi-ethnographic investigation of workers at a nuclear fuel reprocessing plant illuminates these issues. Wynne found that workers at this highly technical and dangerous job, a job that is intimately related to nuclear science and engineering, "defended their ignorance vigorously" (Wynne 1992, p. 39). Wynne discovered their hostility was tied to the perceived disruption of the "necessary social fabric of interdependency" (Wynne 1992, p. 39). As one respondent had explained, knowing how alpha, beta, and gamma radiation works would be unsettling, and thus would create unease among workers and render their work unviable, perhaps even unsafe. The workers in this nuclear fuel reprocessing plant *needed* to be ignorant of the science to do their jobs properly.

These workers depended on the procedures developed by engineers and scientists to keep themselves and each other safe in a potentially dangerous work environment. Reflecting on why these procedures were in place could compromise these efforts. Conceiving of this group as needing more science education misses their lived ecology, and most surveys could not capture this subtle explanation for a public's lack of science literacy. As Jean-Marc Lévy-Leblond argues, the results of studies like Durant, Evans, and Thomas's reveal more about the "inadequacy of the questioning polls than of the illiteracy of the public," largely because people "are much better at answering questions they ask themselves in their professional, political and sentimental lives than they are at answering arbitrary and irrelevant questions asked by outsiders" (Lévy-Leblond 1992, p. 19). In other words, people are better at answering questions when they are relevant and applicable to their organizational, political, or personal ecologies.

Other scholars in public engagement with science have attempted to justify, modify, or outright dispute Durant, Evans, and Thomas's basic defense of the deficit model. Jon Miller (1998) updated their methods by including data from a more open-ended survey conducted in the US in 1995, which he utilized to add an extra dimension of analysis to their study. Steve Miller (2001) noted how the deficit model had begun to lose credibility in the eyes of many public engagement with science researchers, after little ground had been made in educating the public on scientific topics. As Steve Miller suggests, at the turn of the millennium a more contextual approach had taken root for many public engagement with science scholars, very much a response to the critiques that Wynne and Lévy-Leblond had made a decade earlier. Rafael Pardo and Félix Calvo's (2002) essay echoes Steve Miller's claim, but pushes scholars to engage with the public on science policy decisions guided, defined by the "3 D's": discussion, debate, and dialogue (p. 156).

Simis et al. (2016) put forth a plausible explanation for why the deficit model persists today, even though research has demonstrated that public communication of science is more complex than the model suggests. First, they argue that since scientists often draw conclusions based on empirical information, they falsely assume that the general public goes through the same rational processes. Second, they suggest that institutional structures in the education of science, technology, engineering, and math (STEM) fields lack any formal training program in public communication. Therefore, this lack of training influences both best practices and how researchers understand and interpret empirical data in science communication. Third, their study suggests that scientists' perception of a deficit model is closely tied to how they conceptualize "the public." Finally, they argue that the knowledge deficit persists because it is a persuasive argument in public policy circles.

We would add to Simis et al.'s list that the deficit model persists because the survey studies this research utilizes have not been critiqued on the basis of their ecological validity. Falsely assuming that the public uses the same rational processes as scientists is an example of misidentifying the ecology in which a given public is operating, an ecology that is not easily captured by survey studies. Their second claim, that scientists lack training in public communication, only exacerbates the first: this misidentification is in part related to a lack of training in public communication, training which might introduce more robust methods than survey studies. The third argument relates to how scientists often utilize an "idealized" version of the "public," a topic we will return to in Chap. 4, a problem aggravated by the weak ecological validity of survey studies. Their fourth claim demonstrates why more "organic" methods need to be a part of public engagement studies: seeking ecological validity has important implications for our understanding of how publics engage with organizations, technologies, and ideas around them, and more robust data can only help in these efforts.

Two additional qualities make survey instruments poor measures of public engagement, knowledge, or opinion. First, people do not live their lives and make choices as individual units, as surveys measure us, but always as interacting elements of larger functioning bodies. We live our lives within social, professional, and familial ecologies, and our decisions, knowledge, and opinions tend to be those of our ecologies, even shifting over the course of a day as we move between the multiple ecologies in which we operate. The knowledge, opinions, and choices of a larger body are not merely aggregates of the individual elements, any more than a person is a mere aggregate of the organs and organisms that make up a physical body. Surveys tend to isolate individuals from their lived selves, the people they are "in the wild." When we study publics as ecologies instead of individuals as isolated units, we find public knowledge to be significantly higher. What you measure matters as much as how you measure it, and surveys simply do not measure publics.

Second, surveys are poor instruments for measuring *why* a public may lack knowledge. It is at minimum exceedingly difficult, if not impossible, to design a survey that can capture the complicated interrelations that people have within ecologies. Surveys do not give researchers access to life as people actually live. When combined with other methods, it is plausible to suggest that surveys could demonstrate strong ecological validity, but without careful consideration of the operant contexts of a given object of study, surveys do not exemplify the features of strong ecological validity.

Media Studies

Media studies generally gather samples of newspaper, television, and/or magazine items on a topic and then analyze their framing, disposition, metaphors, or similar rhetorical elements to build up a representation of the information reaching a public. At best, media studies can demonstrate stronger ecological validity if the research questions examine media portrayals of the topic, but fall short at enhancing our understanding of actually existing publics. Instead, media studies usually only reveal the dispositions and behaviors of journalists or editors in the production of these stories (Wilkinson et al. 2007).

What researchers interpret from media portrayals or what journalists intend may not reveal how publics actually view or deploy these messages in their respective ecologies. Hence, such studies have weak or strong ecological validity with regards to their descriptions of actual public texts rather than artificial or experimental manipulation of texts. However, they do not have strong ecological validity in relation to any actual public in its reception of those texts. As John Fiske's (1989) works demonstrate, publics treat messages as polysemic, holding a wide variety of sometimes contradictory meanings and finding relevance and utility in a message's adaptability to discourses or practices constitutive of that public (pp. 5–6). Without observing how actually existing publics receive, interpret, deploy, and circulate texts, claims about the relationship of those texts to public understanding or opinion rely upon assumptions that not only are unfounded, but empirically demonstrated to be unsound.

Media studies about climate change have sometimes overlooked these ecological concerns. Craig Trumbo's (1996) content analysis of climate change coverage in five national newspapers reflects how reliance upon theories that standardized or simplify publics' interpretation of texts can weaken the ecological validity of a study. While presenting a thorough analysis of the news story's framing of climate change, Trumbo (1996) relies on Downs's issue-attention cycle as an analytical tool (p. 274). The issue-attention cycle essentially codifies how publics react to public concerns, and by using it as an inference for understanding media framing, Trumbo effectively discounts the impact of ecology, reducing the potential ecological validity of his study. Another example is S. Holly Stocking and Lisa Holstein's (2009) analysis of journalists' roles in the construction of ignorance. Again, their study is a sophisticated and insightful analysis of science journalism, but their reliance on a theory of "manufactured doubt" oversimplifies publics and discounts the impact of ecologies, producing research with weaker ecological validity. For instance, they consider it a "likelihood" that "as the strategic use of ignorance claims to manufacture doubt in scientific controversies grows, public misunderstanding of important scientific issues may be expected to accelerate." (Stocking and Holstein 2009, p. 37). This infers the reality of a public based on presumed reception and use of media, an assumption not accounted for given their data. The truth of their claim is not in question, but its weak ecological validity, its failure to capture how a public actually exists, uses, and modifies the information they encounter, forces them to rely upon shaky assumptions to reach their conclusion.

On the other hand, there are some media studies that proficiently focus on the ecology of science journalism. Peter Weingart, Anita Engels, and Petra Pansegrau's (2000) analysis of climate change discourse in science, politics, and the mass media is a robust example, which achieves a stronger ecological validity than many media studies by circumscribing its scope and comparing the ecology of policy-making, media communication, and scientific discourse. Doing so allows them to circumvent the problems found by making inferences about the public, and achieves sophistication by focusing on science journalism practices. Most importantly, and distinguishing it from studies such as Trumbo's or Stocking and Holstein's is how well Weingart, Engels, and Pansegrau limit their conclusions to those supported by content analysis: claims about the discourse and discourse ecologies analyzed.

Research that utilizes a survey or media study can vary in its level of ecological validity depending on its structure and object of analysis. It may be true that the public lacks knowledge of scientific facts, but a standalone survey study only captures a narrow vision of how a public interacts with scientific information in their everyday lives, and therefore demonstrates weak ecological validity. On the other hand, media studies may be useful for understanding the ecology of science journalism, but inferring public behaviors based on media messages shows weak ecological validity, since these studies usually fail to fully analyze how people encounter, discuss, or understand media messages in their everyday lives. Both methods are common forms of engagement research, but their potential to capture actually existing ecologies is lacking. Oral, curatorial, and digital engagements, however, put scientists and publics in closer proximity, making them a valuable site for thinking about ecological validity.

ORAL, CURATORIAL, AND DIGITAL ENGAGEMENTS

In survey and media studies, communication usually flows in one direction, from the sponsor to the public or from the public to the researcher, which limits the possibilities of fully capturing the ecological relationship between the public and science. Oral and curatorial engagements can also demonstrate these problems; however, the potential for these methods to engage with the public lend themselves to research that can offer stronger ecological validity. Sadly, very little research on science fairs, cafés, or museums has been done by public engagement with science scholars, and digital methods of engagement are relatively new and therefore still lack a large body of scholarship. Therefore, we focus here on a few examples and provide suggestions and directions for future research.

Science Fairs

In the UK, science fairs/festivals can be traced to the British Association for the Advancement of Science's annual conference, which was founded in 1831 to induce discussion between scientists and other educated men of the time. By the 1980s, the conference was renamed the Festival of Science, and is now called the British Science Festival. Many of these festivals are temporary events composed of provisional exhibits, activities, arts organizations, and publics. These festivals are often managed by science museums, universities, independent nonprofit organizations, research councils, or government-funded agencies. The largest science festivals attract between 6000 and 50,000 visitors (Jensen and Buckley 2014, p. 559).

In America, science exhibitions can be traced back to 1828 in New York, with the Science and Technology exposition held by the American Institute of Science and Technology. Technologies like Morse's telegraph and Bell's telephone were exhibited at these showings, and an important aspect of these events was the awarding of medals to distinguished work. By 1928, the focus began to shift towards children, and the 1928 fair, cosponsored by the American Museum of Natural History, is regarded as the first student science fair and the model for all subsequent fairs (Bellipanni and Lilly 1999, pp. 46–47). For the past six years, a children's science fair was even held by President Obama in the White House.

Despite this long history of bringing science to the public, we have very few public engagement with science studies of these events, and those that do exist lack much analysis beyond outputs and attendance. In addition, very few of these studies have focused on actual visitor views or experiences. Those that do study visitor views tend to focus on specific individual events and lack an assessment of the science festival as a whole (Jensen and Buckley 2014, p. 561). When evaluations have been reported, they usually only use closed-ended, self-reported response questions which limits validity. These festivals present a yet-untapped source of understanding how publics interpret, examine, and encounter science in a semi-formal/temporary setting.

Studies that simply examine the output, attendance, or a narrow set of events that occur during the fair will demonstrate weak ecological validity, since they do not capture the actually occurring experience of going to the fair. However, studies like Eric Jensen and Nicola Buckley's (2014) examination of science festivals are an excellent example of research on this topic that makes a greater effort to capture how people actually respond to a science festival. Their approach included an on-site survey that captured the public's opinion at the time, an extended web-based survey that captured the public's opinion soon after the event, as well as focus groups that occurred seven weeks after the event that allowed participants to truly reflect on the experience. We believe their research objectives would be better served by capturing the events of the science festival as they organically unfold in their ecological particularity, and especially by using observational methods to see how people engage with science among peers. While their use of focus groups affords stronger ecological validity than if they had relied upon surveys alone, this research method introduces too many artificial elements to produce valid images of public understanding or opinion. Focus groups tend to be artificial ecologies and, while this can be mitigated by using preexisting groups, the interactions and statements made within a focus group setting are likely different than those made within the participants' ecologies outside the research setting. While not as damaging to ecological validity as many other methods, if a focus group is not composed of a preexisting communicative ecology, its ecological validity will still be weak. Many of the concerns we raise about deliberative engagement events later in this chapter apply equally well to focus groups. Yet, we do find significant value in focus group research and believe the suggestions we offer for organizing organic public engagement events in Chaps. 6 and 7 can easily carry over to modifying focus groups methodologies.

Science Cafés

Science cafés are like public lectures in that scientists and the public convene and participate in discussions about science. However, science cafés are more informal than their public lecture counterparts and often take place in more relaxed environments than lecture halls. Science cafés are becoming increasingly popular in both the UK and The Netherlands as a means to encourage dialogue on a variety of technological and scientific topics between members of the public and stakeholders (Dijkstra and Critchley 2016, p. 73). In contrast to public lectures, science cafés are often organized by volunteers, and these events have seldom been analyzed by public engagement with science scholars.

Anne M. Dijkstra and Christine R. Critchley's (2016) essay is a good example of how these relatively untapped sites of public engagement can be studied. They used a mixed set of methods which included both quantitative and qualitative measures. First, they used a questionnaire to measure respondents' risk perceptions towards nanotechnology. On its own, a study that simply used this method would lack strong ecological validity, because it does not capture the actual process of participating in a science café. However, Dijkstra and Critchley (2016) supplement this method by verbally transcribing the part of the café when speakers gave their presentation and when participants could ask questions (p. 74). In addition, they used a contextual and iterative analysis rather than a frequency analysis, which better captures the actually occurring events of the presentation and therefore produces stronger ecological validity.

We should note that while Dijkstra and Critchley show some success capturing the actual events of the meeting, there are two ways in which the ecological validity of their study could be made stronger. First, they could have utilized both video and audio analysis instead of focusing solely on audio. This would allow them to capture the layout of the room, the way in which participants organized themselves spatially, and so on, and would facilitate analysis of more of the ecology at hand. Alternatively, instead of recording only a few sections of the meeting, they could have recorded more (if not the whole) event. Additionally, while they collected quality observational data about the science café and the activities of participants at the event, they relied solely on a questionnaire for comparative data about individuals who did not attend.

Science Museums

Science museums can be traced back to the turn of the seventeenth century, when specimen collections, once reserved for royalty and aristocrats, were opened to the public. By the eighteenth century, scholars began to take pains to gather and preserve natural and man-made samples. Henceforth, the functions of the museum became to "show what is worthy of being conserved in order to educate" (Schiele 2008, p. 28). The mission of the museum is very much connected with public engagement with science's emphasis on the deficit model, which positions this curatorial experience as a plausible means to overcome the public's misunderstanding of science.

There has been a large amount of research on science and technology museums spanning journals in both public engagement with science and museum studies. These range in their approach and research questions, including views of museum professionals (Henriksen and Frøyland 2000), the interactivity of the museum (Yaneva et al. 2009), the art-culture divide (Shein et al. 2015), and expectations and experiences of attendees and curators (Kamolpattana et al. 2015). Their methods are equally varied, and many attempt to measure museum-goer's experiences with surveys, questionnaires, focus groups, or direct observations. Studies that focus on how attendees interact with the museum will be having stronger ecological validity than those that simply measure their attitudes before, during, or after their visit. In turn (and much like studies of science cafés), studies that examine museums can increase their ecological validity by increasing the scope of what they analyze at the museum, including a variety of different exhibits and interactions that occur between visitors, the curators, and the spatial distribution and layout of the museum itself.

The Association of Science-Technology Centers' overview of studies on museums notes that gaps in research on science museums include assessments and descriptions of the communicative activities of science museums and analysis of what kinds of images of science are presented/ communicated to the public. Taking account of the ecological validity of research on museums can help fill some of these gaps, and studies like Albena Yaneva, Tania Mara Rabesandratana, and Birgit Greiner's (2009) analysis of interactivity (which uses quasi-ethnographic qualitative methods) show promise for taking account of ecological validity in these important curatorial engagement spaces. Their study was openly experimental, created a wholly new ecology, and took as part of its task the study of how people would engage that new ecology. In this sense, the ecology is "artificial" and experimental and the study has no capacity to speak to questions of science communication or public engagement with science outside of the contrived setting. At the same time, the freedom given to participants to engage on their own terms and in their own ways, combined with the open and observational mode of data collection, means that at least what they conclude about how publics might interact with such a contrived space can be afforded higher ecological validity.

Digital Spaces

The popularization of personal computers and the internet has had a profound influence on all fields of research, public engagement with science included. Scientists are socialized into internet culture at early points in their careers, and communicating via the internet has become a natural part of their personal and professional lives. The internet functions as a method of collaboration, a space for the facilitation of competition, and a locale where disciplinary subspecialists can network through email and online discussion groups. The ubiquity of digital and online practices of scientists is observable in efforts to engage the public as well. Most notably, the deficit model has found traction in research on online engagements. As Michael Cacciatore and others have argued, the internet may help "provide a more appropriate backdrop for in-depth discussions...into controversial areas, thereby enhancing the likelihood of narrowing gaps" (Cacciatore et al. 2014, p. 389).

One example of how scientists engage the public digitally is the trend toward open-access journals, which have "significant implications for public communication of science and technology" (Trench 2008, p. 187). An initiative to open scholarly research online is an application of communalist norms, and the Human Genome Project's commitment to public chromosome sequences is an example of this application. Alternatively, the Public Library of Science (PLoS) has launched its own open-access journals on medicine, biology, genetics, pathogens, and neglected tropical diseases. While significant, examining how publics actually utilize these journals would be impractical at best. Therefore, as a potential site for research with strong ecological validity, examining how publics engage science with open-access journals falls short.

Science blogs have also become popular spaces of online engagement between science journalists and the public. Unlike traditional media studies, the online format allows researchers to get a more robust representation of how publics actually engage with scientific topics. Mathieu Ranger and Karen Bultitude's (2016) study on bloggers' practices relating to audience recruitment is a sophisticated example of this kind of research. Ranger and Bultitude (2016) measured blog popularity, performed content analysis, and conducted semi-structured interviews. Their analysis concluded that science journalists take many factors into account to attract visitors to their blogs (pp. 374–375). While their study primarily focuses on the ecology of science journalism and therefore avoids many of the missteps possible with traditional media studies, Ranger and Bultitude ignore some facets of the internet blogging that could infuse their study with stronger ecological validity. For instance, analyzing how readers respond in the comments section, or how any dialogue emerges organically from this discussion will strengthen an online study's potential ecological validity.[1]

Online forums have limited value for developing studies with strong ecological validity, but ecological validity can be maximized in these cases if researchers pay close attention to how engagement organically unfolds within the digital space under investigation. Research by both Carcioppolo et al. and Christine Hine analyzed online public forums as sites of public engagement. Carcioppolo et al.'s (2016) study looked at in-group rationalizations of risk of indoor tanning practices, and thematically categorized responses according to the kinds of rationales forum users utilized. While looking at online forums can be a productive mode of developing a study with ecological validity in mind, this study limits its potential ecological validity by only examining and categorizing the kinds of messages users were using. Instead, looking at how discussion occurs *between* users might have given researchers a better understanding of how the public actually engages online with a scientific topic. On the other hand, Hine's (2014) study of online discussions of head lice is highly influenced by calls for contextual public engagement methods from scholars like Irwin and Wynne. While Hine's study is limited because of ethical concerns over revealing users' personal information, her qualitative analysis more than accounts for this problem by capturing the ecology of an everyday contextual exchange of discourse between users. Hine's work is often very attentive to questions of ecology and ecological validity, and we will return to it in greater detail in Chap. 6.

Citizen science is another developing field of public engagement with science and can best be described as the collective online work of publics on scientific tasks often under the direction of professional scientists. One popular example of citizen science is an online platform called "Zooniverse," run by the Citizen Science Alliance. There are currently around 1.5 million registered users, and projects on Zooniverse range from genetics to astronomy. Karen Masters (2016) and her fellow researchers analyzed whether this example of citizen science developed participants' scientific literacy. Their research provides a sense of how this new online platform might fit into the public engagement with science paradigm, but their sole reliance on a survey (and an accompanying quiz) lends their preliminary research weak ecological validity (Masters et al. 2016, pp. 5–9). It is difficult to say how popular citizen science will be and for how long, but this new form of public engagement could be aided by the inclusion of research using observation and interviews with participants, designed with attention to preserving the ecologies being studied.

DELIBERATIVE ENGAGEMENTS

Neither surveys nor media studies dominate discussions of public engagement, and research on oral, curatorial, and digital spaces of public engagement is still rare. Since at least the 1970s, the term "public engagement" has most commonly evoked a model that brings people together, usually with experts and policy-makers, to have a discussion that follows a relatively narrow set of norms ranging from consensus conferences and citizen panels to citizen juries and deliberative polling. In this section, we examine these increasingly popular modes of public engagement and assess their potential ecological validity.

Most deliberative engagement methodologies share an experimental design caught between analytic research methods and idealist philosophies of public engagement. Both imperatives, which drive the majority of public engagement events, push researchers and organizers toward increasingly artificial contexts. Experimental studies of public engagement, such as Michael Cobb's (2005) study of framing effects on public opinions of nanotechnology, tend to use classical experimental design, build artificial populations, study them as individuals rather than ecologies, attempt to draw conclusions by aggregating from individual behaviors, and manipulate variables to look for effects. Such approaches are motivated in part by a desire for, as Jane Macoubrie (2006) put it, "efficiency and control" (p. 223). Efficiency is prioritized because of limited time and resources. Control is desirable because sufficient control over the experiment assures some viable research outcome that will lead to publication and satisfy granting agencies. Surrendering control to actually existing publics risks running a study only to discover the presumptive justification for the study was unfounded. Additionally, some researchers have a very limited set of methods and need to control the research design to ensure the applicability of those methods. Rather than build their methods out of the demands of the object of inquiry, they control the object of inquiry to fit their methods. Some may even see every question as one naturally fitted to their limited tools. Known as Kaplan's Law of Instrument (Kaplan 1964, pp. 28–29), this problem was best summarized by Maslow (1966) in the pithy line, "I suppose it is tempting, if the only tool you have is a hammer, to treat everything as if it were a nail" (p. 16).

In addition to the limitations of analytic methods, an idealist vision further impedes research by presuming that existing publics do not already deploy good deliberative methods and must be trained or facilitated to have a meaningful engagement. One premise of such engagement events is that existing publics are insufficiently diverse. Thus, because of both sampling methods imposed by analytic methods and a presupposition that people ought to speak to others outside their conventional encounters, researchers build engagement events by cross-sampling a population and selecting individuals across demographic characteristics. For example, Daniel Merkle (1996) reported that a large-scale national deliberative poll randomly selected participants, uprooted them from their cultural, social, and even geographic ecologies, and then placed them into an alien environment and into conversation with people with whom they had no previous experience. Alternatively, Caroline Lee (2016) described how organizers of the Community Congress deliberative event in post-Katrina New Orleans struggled to produce an "accurate" racial representation of the city, resulting in an engagement with "a disproportionate number of white returnees in elevated areas of the city" (Lee 2016, p. 14). These kinds of sampling problems reflect the divide between the ideal representative sample deliberative researchers strive to create and the actual ways in which publics organize, coalesce, divide, and move. In short, the sampling method creates artificial aggregates of people who would never actually form a public.

Even if the intention of such an event is only to give non-experts the opportunity to endorse, criticize, or prioritize potential policy options, the findings on ecological validity in jury research and work such as Epstein's (2000) study of AIDS treatment activists make clear that problems of setting, sampling, mode of deliberation, and perceived consequences complicate the validity of the results. People form different ideas, process information differently, and reach different conclusions when uprooted from their ecologies and placed into artificial settings, instructed to deliberate in unfamiliar ways, and perceive different consequences. As we will discuss in greater detail in Chap. 4, the philosophies foundational to most deliberative engagements presume that publics must be altered to merit participation in the policy process, believing that publics as they exist will reach conclusions unworthy of inclusion. Thus, the members of publics brought into such events no longer represent any existing public and, hence, cannot provide decision-makers with the voices of actual publics. Instead, the counterfactual publics constructed by such engagement methods primarily reflect the engagement expert's values and beliefs about deliberation and policy-making.

Even many studies that claim ecological validity or are only "quasi-experimental" tend to build counterfactual research spaces. Erik Nisbet's (2006) study of opinion leadership, for example, considered ecological validity only in the sense of cultural differences between nations or countries, but failed to consider how peculiar communicative and social ecologies can be. Likewise, Macoubrie (2006) minimized the number of variables being altered by the researchers so that certain items could form organically, but still operated in an almost entirely artificial laboratory environment, without acknowledging how setting affects communication, deliberation, and engagement.

Deliberative democracy idealists analyze and utilize many different modes of deliberative engagement. In the rest of this section, we examine consensus conferences, citizen panels, citizen juries, and deliberative polls. For each of these different methods, we provide examples and substantiate our claims about the idealism that surrounds these increasingly popular engagement methods.

Consensus Conferences

Originally created in the US but modified by the Danish Board of Technology in 1987 for public engagement purposes, the consensus conference is a tool of participatory assessment that brings together citizens and experts for the sake of exchanging ideas and developing a consensus decision based on dialogue. Generally, these conferences are composed of a panel of ten to twenty citizens recruited through invitations sent to a large random sample. Respondents to the call are then selected based upon their demographics. In addition, an expert panel is developed to engage in conversation with the selected citizens. There is variance in how consensus conferences are run, but the central element of consensus remains consistent across examples. Studies that have utilized consensus conferences in the past few decades have centered around the topic of technology or biotechnology.

Einsiedel et al.'s (2001) study is an early example of public engagement, with science scholars investigating the potential of consensus conferences for gauging public engagement with genetically modified organisms. They looked at how well consensus conferences "traveled" across a number of different contexts, including Denmark, Canada, and Australia. Their analysis concludes that consensus conferences hold value in these contexts, and their presence in public engagement practices lends legitimacy to groups interested in furthering technological dialogues. However, their study never mentions the details of how these conferences unfolded, and instead gives a broad analysis of similarities between events in different countries. This makes it impossible to assess the ecological validity of their research, though all indications are that no concern was given to the role of ecology or its impacts on research design. This ellipsis of a fundamental form of validity for all research, and one that is especially salient to public engagement research, is all too common and a substantial reason why so much public engagement research yields no useful or actionable knowledge.

Joanna Goven's (2003) analysis, however, presents a more thorough account of the proceedings of a consensus conference on plant biotechnology held in New Zealand in 1996. Goven's study explicitly recognizes the problematic idealist stance when she refers to how the organizers of the event attempted to mitigate their bias. The organizing committee for the event was friendly towards the technology under deliberation, and sought to "manage the possible resulting bias by ensuring that there were oppositional views expressed at the conference" (Goven 2003, p. 428). As Goven explains, participants were inundated with information on plant biotechnology with little time for reflection, the conferences themselves devoted little time to allow panelists to discuss issues among themselves, and the way the conference was established made it appear as if there was little to no disagreement among the scientists (the only people granted "expert" status by the study). Making sure that opposition occurs in the deliberative process between organizers and the public imposes an artificial argumentative space on the participants. Additionally, considering the artificial time constraints levied upon panelists, this deliberative event is merely counterfactual. Hence, any inferences drawn or claims developed out of this engagement process will possess weak ecological validity. Given the overwhelming research indicating that changes in method of deliberation affect outcomes, the conclusions we can draw from Goven's study are strictly limited to the artificial space and model created by the structured event.

More recently, researchers Ashley Anderson et al. (2012) have begun to investigate the "informal" aspects of consensus conferences, and include a critique of deliberative idealism similar to the one we make in Chap. 4. As they put it, these "critiques are ironic, if not tragic, given that a primary purpose of public engagement exercises is to empower lay citizens to impact the development of science and technology" (Anderson et al. 2012, p. 956). As these researchers note, one way to relieve the problems of deliberative idealism is to incorporate elements of information-seeking by participants. During these information-seeking meetings, facilitators ask panelists to engage in their own information-seeking and keep a journal of their findings. Anderson and her fellow researchers rightly argue that incorporating a structured element of information-seeking to a consensus conference can strengthen the ecological validity of such research. However, despite attempts to fold informal elements into the deliberative process, the fact remains that plucking panelists out of their native ecologies, disrupting the settings and populations in which they usually participate, will always prevent deliberative events from making ecologically valid claims about behaviors, knowledge, or opinions as they occur outside the contrived setting of the consensus conference.

Citizen Panels

Much like consensus conferences, citizen panels function to put experts and citizens in dialogue. However, citizen panels are not as singularly focused on consensus. Instead, an average citizen panel consists of a group of lay citizens who meet for a few days, learn or discuss one or more political issues, confer with an expert panel, and then hold a press conference to publicize the results. Panelists are given equal time to speak, and cooperation is valued over self-interestedness (Brown 2006, p. 204). To ensure this type of participation, panelists are sometimes provided a professional facilitator and clerical staff. Organizers of the event (usually trained facilitators, engagement experts, researchers, or event sponsors) often seek consensus, but will allow minority reports when consensus proves too difficult. These panels have gained traction on topics as diverse as child welfare, biotechnology, and environmental conservation.

In the US, a 1996 amendment to the Child Abuse Prevention and Treatment Act called for the development and implementation of a citizen review panel in each state by 1999. These panels consist of a representative sample of the community, meet quarterly, and submit an annual report to the federal government outlining their activities and recommendations. Furthermore, panelists review the practices and policies of the child protection system, and each state child welfare agency must respond in writing to the panel's recommendations. Blake Jones and David Royse's (2008) review of literature on the subject revealed that these public engagement events had little associated scholarship (p. 145). The small body of research that does exist is often focused on how they are designed, function, or are viewed by the panelists themselves. Conspicuously absent is any exploration of how effective the panelists' suggestions are for policy decisions or their presence and affect in the communities they serve. Despite their potential importance, this instantiation of a citizen panel possesses weak ecological validity because it creates an artificial environment where community members are asked to perform tasks removed from their already-existing ecologies. The lack of research on these panels makes it difficult to assess the full extent of their ecological validity, but from the little data that does exist, it is plausible to suggest these forms of public engagement do not capture an existing ecology. In fact, as we will discuss further in Chap. 4, such structures remove the panelists from their role as citizens and transform them into a kind of civil servant.

An example of citizen panels focused on emerging technology is Johan Evers and Joel D'Silva's (2009) analysis of the "Bio-on-Chip" technology held by the Flemish participatory Technology assessment project. This panel consisted of 15 participants who were asked to reflect on two hypothetical future situations centered around nanotechnology. These thought experiments were not designed for predictive value, but instead used to gain a picture of what concerns the public might have for future directions of nanotechnology. In order to facilitate this hypothetical situation, two scenarios were invented consisting of a simple plot and a main character. In workshops, these main characters were played by professional actors and panelists in role-playing exercises. Much like similar studies on deliberative processes, Evers and D'Silva (2009) construct a typology for organizing and categorizing panelists' discourse, and draw conclusions about the ethical values at play in nanotechnology research, noting that moral reasoning about any subject will be context-dependent (p. 139). The ecological validity of their conclusions is suspect due to the artificiality of the setting under examination (especially the use of paid actors and science fiction narratives), which does not replicate the conditions under which panelists would deliberate about the technologies in question in their actually existing ecologies. In addition, the fictive narrative told to participants alters the perceived outcomes of the event; since panelists knew that the scenarios were hypothetical, their deliberative process removed the stakes of the decision-making process and thus also failed to capture the actually existing ecology.

Finally, Robinson et al.'s (2014) study of the Responsible Science and Public engagement workshop served to introduce panelists to the importance of environmental conservation. Panelists were given a tour of a local woodland, provided a "handcrafted booklet" of artwork and historical maps of the area, and made to perform a series of "performative tasks" in the woodlands designed to encourage participants to form bonds of mutual respect (Robinson et al. 2014, p. 84). A series of artistic performances and activities ensued, all led by artists involved in the project. Robinson et al. suggest that these activities in nature provided a "Heideggerian experience of *Dasein*, 'Being-in-the-world'," by positioning participants in a lived reality where they touched, felt, and smelled the environment around them

(Robinson et al. 2014, p. 84). Robinson et al. disregard one of the central features of *Dasein*: its "thrown-ness" in the world or, how all beings are thrown into an already-existing ecology which colors their understanding of the world and others. The artificiality of their engagement event does not reflect how participants exist in their own meaningful ecologies, or how they are "thrown" into their own ecologies, and therefore lacks the capacity to capture how the public "dwells" in their constitutive environments. Claims about publics that emerge from these kinds of events, in turn, will possess weak ecological validity.

Citizen Juries

Citizen Juries are a deliberative process developed in the 1970s by Ned Crosby, who founded the Jefferson Center for New Democratic Processes. Just like consensus conferences and citizen panels, citizen juries are constructed by selecting participants that resemble the community. These juries are composed of 24 people who are given information from experts and advocates presenting different points of view. Moderators attempt to facilitate discussion, not lead the group toward any preconceived conclusion and attempt to minimize the presence of their own biases. Additionally, a citizen jury event usually lasts three to five days and at the end of it the members present their recommendations to the organizers. Examples of citizen juries can be found spanning numerous topics, including healthcare and public policy.

King et al.'s study using citizen juries examined public perspectives on technologies that enable new kinds of information and communication exchange in healthcare settings (e-health) (King et al. 2010, p. 352). Researchers in this study conducted two juries: one urban and one rural, and attempted to recruit members based on a "balanced" subset of demographics. Prior to the jury, participants were given some introductory materials, a schedule of the day, the aims of the jury, and a set of questions to be debated. The first half of the day was reserved for educational purposes, and jury members watched a DVD overviewing e-health and heard testimony from clinicians with opposing perspectives. Oddly enough, jury members were not required to use e-health applications or software in order to form their opinions, hence, the deliberative process in this case in no way resembles how these citizens would actually encounter e-health applications. King et al.'s (2010) conclusion that jurors "agreed unanimously that e-health should be developed and was in many ways inevitable," while perhaps true in this counterfactual scenario, is derived from a research design with very weak ecological validity and is therefore impossible to extrapolate to other publics (p. 355).

Another example is Jefferson Action's "Reclaim November Ohio," which consisted of two citizen juries. The first jury lasted three days, and members were asked to set the agenda for the second jury. In this first engagement event, the jury was presented with ten issues broadly related to the economy and listened to expert testimony until they agreed on what should be addressed in the subsequent jury. The second jury participated in two separate events, lasting three and four days respectively. Juries during this event listened to expert testimony on the items decided by the first jury and then developed questions for "congressional candidates and interviewed both the candidates and their campaign staff about their positions on, and plans for, addressing the three economic issues" (Munno and Nabatchi 2014, p. 8). Greg Munno and Tina Nabatchi (2014), researched these events in the hope of discovering whether deliberation offers agency and voice to citizens, and whether these events change people's perceptions of politics in the US. While the setup of this jury was complex and sophisticated and while Munno and Nabatchi's analysis was thorough and detailed, this event is clearly a counterfactual, where political policies and ideas are artificially deliberated on by a carefully curated set of participants gathered in a contrived setting. Conclusions about how these deliberative engagements might help the democratic cause, like Munno and Nabatchi have done, possess weak ecological validity because they lack the organic structure of deliberation found in actually existing public ecologies. We may be able to make some conclusions about how a curated selection of individuals operates in this specific contrived setting from their research, but we cannot extrapolate from that data to how actually existing publics function.

Street et al.'s (2014) systematic review of the use of citizen juries for health policy decision-making is one of the most comprehensive meta-studies done on this topic, and provides a clear picture of this engagement method's ecological validity across several studies. Street and her fellow researchers narrowed down the body of health policy articles that use deliberative methods to 37 papers, which describe 66 citizen juries or variants thereof. As we have emphasized in this chapter, facilitation by organizers will inevitably lead to weaker ecological validity in subsequent analyses, because publics do not deliberate through facilitation by experts in their actually existing ecologies. Of the 37 studies under examination, only 3 were not facilitated (Street et al. 2014, p. 2). Furthermore, despite hopes that deliberative methods such as citizen juries can reinvigorate democratic ideals, only three studies investigated by Street et al. described "commitment by a decision-making body to consider recommendations" (Street et al. 2014, p. 2) Even taken as a counterfactual, most citizen juries are purely simulated engagement events that do not capture public opinions and whose results have little to no bearing on actual decision-making processes.

Deliberative Polls

Deliberative polling is a method developed by James Fishkin at the Center for Deliberative Democracy housed at Stanford University. Deliberative polling begins by sending out a questionnaire to a random and representative sample of the public. Of this sample, a smaller subset of participants is randomly selected to participate in the deliberative poll. In contrast to citizen juries and consensus conferences, where participants are hand-picked, deliberative polling makes the point of including a random subset of participants (Fishkin et al. 2000, p. 660). Prior to the deliberative event, participants are given briefing materials on the topics to be discussed, and when they arrive at the event participants are randomly assigned to small groups that include trained moderators. These groups develop and pose questions to experts and policy-makers, and the event concludes by having the participants engage in a final questionnaire. The results of this final survey are then analyzed and released to the media.

Francis Schweigert's (2010) analysis of an engagement event on transportation funding and priorities is a representative example of deliberative polling in practice. Unlike citizen juries, citizen panels, or consensus conferences, deliberative polling more closely resembles an advanced or elaborate method of performing a survey study than it does a proper engagement event. Much of the data gathered from these deliberative polling events has little to do with decision-making, and instead measures how much participants' views changed after engaging in group discussion. Schweigert adeptly gathers and relays the findings from this event, but perhaps more interestingly (and unlike many studies on deliberative events), addresses some potential critiques of deliberative polling methods.

The first critique asks if the process of deliberative polling can "clarify what people truly value, or does it distort through hidden biases and the pressures of group interaction?" (Schweigert 2010, p. 31). As Schweigert notes, one way to investigate this critique is to determine if the opinions expressed are more in line with their background beliefs, or show greater affinity with the expected consequence of the policy under investigation. Research seems to suggest the internal consistency of participant views has not been confirmed (Schweigert 2010, p. 32). The second critique submits that the group dynamic of the deliberative process can alter participant's views artificially. One possible cause for this artificial change would be hidden biases in small group discussions that would leave less rhetorically gifted participants at the mercy of those with greater social skills. A second cause is a shift in group thinking merely as a result of group interaction, better known as the "Hawthorne Effect" (treatment group performs better while being observed).

Schweigert argues that both problems are countered by deliberative polling design. The first critique is foiled by the moderation of trained facilitators who make sure "discussions are explicitly not directed toward agreement on views" (Schweigert 2010, p. 32). The second critique is stymied by the way in which deliberative polling seeks to "change the environment and quality of interaction, rather than to attempt to mimic the less engaged status quo" (Schweigert 2010, p. 33). Ironically, both defenses of deliberative polling confess to weak ecological validity in research design. In the first case, distortions in deliberative processes will always occur when you remove participants from their already-existing ecologies. Adding in a moderator does not alleviate this problem; in fact, it only serves to increase the artificiality of these events. It does not matter if discussions are directed towards consensus or disagreement; the fact that they are directed at all is problematic for a study's ecological validity. In the second case, Schweigert is correct in suggesting that deliberative polling seeks a different end than the "less engaged status quo." However, claims about how the public changes its opinion, how they deliberate in small groups, or even how they ask questions in their actually existing ecologies cannot be derived from research on deliberative polling. The only data with strong ecological validity that comes from these studies is data about what deliberative polling does to the participants involved in deliberative polling. Claims about how the public thinks, acts, or otherwise deliberates will have little to no ecological validity, since the data gathered does not capture actually occurring ecologies of engagement.

As Darrin Hicks (2002) and Darrin Durant (2011) have noted, the idealized vision of public deliberation underlying much of engagement research is derived from the philosophies of John Rawls and Jürgen Habermas. While Durant notes significant differences between Rawls and Habermas, both present political philosophies that profess standards for "good" deliberation that engagement researchers recognize as counterfactual, yet normative. Seong-Jae Min's (2007) study of computer-mediated engagement articulates the norms that pervade most public engagement events: researchers self-consciously manipulate the publics being studied through instructions, mediation, or facilitation to reflect and enforce "assumptions and criteria of deliberation—equality, rationality, and civility of communication" (pp. 1381–1382). Hence, the public, in most public engagement, not only is an artificial construct (a group unlikely to gather organically) but is furthermore instructed or facilitated to deliberate differently than how they would outside the experiment. As a result, what we can conclude with confidence about the majority of public engagement events is that they study the potentialities for the researcher's vision of a utopian public. Such research has ecological validity only if one is studying that artificial and counterfactual ideal, but cannot provide a basis for extrapolating to actually existing publics.

Ecologically valid public engagement engages actual publics and observes them in their existing ecologies. It considers the broad range of questions about setting, population, relationships, communication patterns, qualities of deliberation, and so on, as they already occur in that particular public. If we develop a body of ecologically valid public engagement research, we may begin to draw a picture of the myriad ways in which diverse publics deal with new information, questions of risk, and other issues of concern. Absent such work, we learn far more about idealist models of public deliberation than about actually existing publics.

NOTES

1. Alternatively, examining a more dialogic forum-based discussion, such as Reddit's "Ask Me Anything" section could provide an interesting potential case study for examination.

REFERENCES

Anderson, Ashley A., Jason Delborne, and Daniel Lee Kleinman. 2012. Information Beyond the Forum: Motivations, Strategies, and Impacts of Citizen Participants Seeking Information During a Consensus Conference. *Public Understanding of Science* 22 (8): 956–970.

Bellipanni, Lawrence J., and James Edward Lilly. 1999. What Have Researchers Been Saying About Science Fairs? *Science and Children* 36 (8): 46–50.

Brown, Mark B. 2006. Survey Article: Citizen Panels and the Concept of Representation. *Journal of Political Philosophy* 14 (2): 203–225.

Cacciatore, Michael A., Dietram A. Scheufele, and Elizabeth A. Corley. 2014. Another (Methodological) Look at Knowledge Gaps and the Internet's Potential for Closing Them. *Public Understanding of Science* 23 (4): 376–394.

Carcioppolo, Nick, Elena V. Chudnovskaya, Andrea Martinez Gonzalez, and Tyler Stephan. 2016. In-group Rationalizations of Risk and Indoor Tanning: A Textual Analysis of an Online Forum. *Public Understanding of Science* 25 (5): 627–636.

Cobb, Michael D. 2005. Framing Effects on Public Opinion About Nanotechnology. *Science Communication* 27 (2): 221–239.

Dijkstra, Anne M., and Christine R. Critchley. 2016. Nanotechnology in Dutch Science Cafés: Public Risk Perceptions Contextualised. *Public Understanding of Science* 25 (1): 71–87.

Durant, Darrin. 2011. Models of Democracy in Social Studies of Science. *Social Studies of Science* 41 (5): 691–714.

Durant, John, Geoffrey Evans, and Geoffrey Thomas. 1992. Public Understanding of Science in Britain: The Role of Medicine in the Popular Representation of Science. *Public Understanding of Science* 1 (2): 161–182.

Einsiedel, Edna F., Erling Jelsøe, and Thomas Breck. 2001. Publics at the Technology Table: The Consensus Conference in Denmark, Canada, and Australia. *Public Understanding of Science* 10 (1): 83–98.

Epstein, Steven. 2000. Democracy, Expertise, and AIDS Treatment Activism. In *Science, Technology, and Democracy*, ed. Daniel Lee Kleinman, 15–32. Albany: State University of New York Press.

Evers, Johan, and Joel D'Silva. 2009. Knowledge Transfer from Citizens' Panels to Regulatory Bodies in the Domain of Nano-Enabled Medical Applications. *Innovation: The European Journal of Social Science Research* 22 (1): 125–142.

Fishkin, James S., Robert C. Luskin, and Roger Jowell. 2000. Deliberative Polling and Public Consultation. *Parliamentary Affairs* 53 (4): 657–666.

Fiske, John. 1989. *Reading the Popular*. London: Routledge.

Goven, Joanna. 2003. Deploying the Consensus Conference in New Zealand: Democracy and De-problematization. *Public Understanding of Science* 12 (4): 423–440.

Henriksen, Ellen K., and Merethe Frøyland. 2000. The Contribution of Museums to Scientific Literacy: Views from Audience and Museum Professionals. *Public Understanding of Science* 9 (4): 393–415.

Hicks, Darrin. 2002. The Promise(s) of Deliberative Democracy. *Rhetoric and Public Affairs* 5 (2): 223–260.

Hine, Christine. 2014. Headlice Eradication as Everyday Engagement with Science: An Analysis of Online Parenting Discussions. *Public Understanding of Science* 23 (5): 574–591.

Jensen, Eric, and Nicola Buckley. 2014. Why People Attend Science Festivals: Interests, Motivations and Self-Reported Benefits of Public Engagement with Research. *Public Understanding of Science* 23 (5): 557–573.

Jones, Blake L., and David Royse. 2008. Citizen Review Panels for Child Protective Services: A National Profile. *Child Welfare* 87 (3): 142–162.

Kamolpattana, Supara, Ganigar Chen, Pichai Sonchaeng, Clare Wilkinson, Neil Willey, and Karen Bultitude. 2015. Thai Visitors' Expectations and Experiences of Explainer Interaction Within a Science Museum Context. *Public Understanding of Science* 24 (1): 69–85.

Kaplan, Abraham. 1964. *The Conduct of Inquiry: Methodology for Behavioral Science*. Rutgers: Transaction Publishers.

King, Gerry, David J. Heaney, David Boddy, Catherine A. O'Donnell, Julia S. Clark, and Frances S. Mair. 2010. Exploring Public Perspectives on E-health: Findings from Two Citizen Juries. *Health Expectations* 14 (4): 351–360.

Lee, Caroline W. 2016. *Do-It-Yourself Democracy*. New York: Oxford University Press.

Lévy-Leblond, Jean-Marc. 1992. About Misunderstandings About Misunderstandings. *Public Understanding of Science* 1 (1): 17–21.

Macoubrie, Jane. 2006. Nanotechnology: Public Concerns, Reasoning and Trust in Government. *Public Understanding of Science* 15 (2): 221–241.

Maslow, Abraham. 1966. *Psychology of Science: A Reconnaissance*. New York: Harper & Row.

Master, Karen, Eun Young, Joe Cox, Brooke Simmons, Chris Lintott, Gary Graham, Anita Greenhill, and Kate Holmes. 2016. Science Learning via Participation in Online Citizen Science. *Journal of Science Communication* 15 (3): 1–33.

Merkle, Daniel M. 1996. The Polls – Review: The National Issues Convention Deliberative Poll. *Public Opinion Quarterly* 60 (4): 588–619.

Miller, Jon D. 1998. The Measurement of Civic Scientific Literacy. *Public Understanding of Science* 7 (1): 203–223.

Miller, Steve. 2001. Public Understanding of Science at the Crossroads. *Public Understanding of Science* 10 (1): 115–120.

Min, Seong-Jae. 2007. Online vs. Face-to-Face Deliberation: Effects on Civic Engagement. *Journal of Computer-Mediated Communication* 12 (4): 1369–1387.

Munno, Greg, and Tina Nabatchi. 2014. Public Deliberation and Co-production in the Political and Electoral Arena: A Citizens' Jury Approach. *Journal of Public Deliberation* 10 (2): 1–29.

Nisbet, Erik C. 2006. The Engagement Model of Opinion Leadership: Testing Validity Within a European Context. *International Journal of Public Opinion Research* 18 (1): 3–30.

Pardo, Rafael, and Félix Calvo. 2002. Attitudes Toward Science Among the European Public: A Methodological Analysis. *Public Understanding of Science* 11 (2): 155–195.

Ranger, Mathieu, and Karen Bultitude. 2016. 'The Kind of Mildly Curious Sort of Science Interested Person Like Me': Science Bloggers' Practices Relating to Audience Recruitment. *Public Understanding of Science* 25 (3): 361–378.

Robinson, Philip A., Phil MacNaghten, Sarah Banks, Janie Bickersmith, Angela Kennedy, Yvonne Richardson, Sue Shaw, and Ingrid Sylvestre. 2014. Responsible Scientists and a Citizens' Panel: New Storylines for Creative Engagement Between Science and the Public. *The Geographical Journal* 180 (1): 83–88.

Rowe, Gene, and Lynn J. Frewer. 2005. A Typology of Public Engagement Mechanisms. *Science, Technology, & Human Values* 30 (2): 251–290.

Schiele, Bernard. 2008. Science Museums and Science Centres. In *Handbook of Public Communication of Science and Technology*, ed. Massimiano Bucchi and Brian Trench, 27–39. New York: Routledge.

Schweigert, Francis J. 2010. Strengthening Citizenship Through Deliberative Polling. *Journal of Community Practice* 18 (1): 19–39.

Shein, Paichi P., Yuh-Yuh Li, and Tai-Chi Huang. 2015. The Four Cultures: Public Engagement with Science Only, Art Only, Neither, or Both Museums. *Public Understanding of Science* 24 (8): 943–956.

Simis, Molly J., Haley Madden, Michael A. Cacciatore, and Sara K. Yeo. 2016. The Lure of Rationality: Why Does the Deficit Model Persist in Science Communication? *Public Understanding of Science* 25 (4): 400–414.

Stocking, S. Holly, and Lisa W. Holstein. 2009. Manufacturing Doubt: Journalists' Roles and the Construction of Ignorance in a Scientific Controversy. *Public Understanding of Science* 18 (1): 24–42.

Street, Jackie, Katherine Duszynski, Stephanie Krawczyk, and Annette Braunack-Mayer. 2014. The Use of Citizens' Juries in Health Policy Decision-Making: A Systematic Review. *Social Science and Medicine* 109: 1–9.

Trench, Brian. 2008. Internet: Turning Science Communication Inside-Out? In *Handbook of Public Communication of Science and Technology*, ed. Massimiano Bucchi and Brian Trench, 185–198. New York: Routledge.

Trumbo, Craig. 1996. Constructing Climate Change: Claims and Frames in US News Coverage of an Environmental Issue. *Public Understanding of Science* 5 (3): 269–283.

Weingart, Peter, Anita Engels, and Petra Pansegrau. 2000. Risks of Communication: Discourses on Climate Change in Science, Politics, and the Mass Media. *Public Understanding of Science* 9 (3): 261–283.

Wilkinson, Clare, Stuart Allan, Alison Anderson, and Alan Petersen. 2007. From Uncertainty to Risk?: Scientific and News Media Portrayals of Nanoparticle Safety. *Health, Risk & Society* 9 (2): 145–157.

Wynne, Brian. 1992. Public Understanding of Science Research: New Horizons or Hall of Mirrors? *Public Understanding of Science* 1 (1): 37–43.

Yaneva, Albena, Tania Mara Rabesandratana, and Birgit Greiner. 2009. Staging Scientific Controversies: A Gallery Test on Science Museums' Interactivity. *Public Understanding of Science* 18 (1): 79–90.

CHAPTER 4

Publics and Counterpublics of Engagement

Unclear goals and problems of ecological validity have produced a legitimacy crisis for public engagement. Richard Jones (2007) notes that public engagement is caught between competing objectives. On the one hand, the cynical purpose of engagement is "to defuse potential public opposition" and "obtain some fig leaf of public consent to a decision that has already been made" (Jones 2007, p. 262). Pidgeon and Rogers-Hayden (2007) echo this criticism, claiming that much public engagement with science only occurs late in research and development, "after a controversial ethical or risk question has arisen" (p. 194). On the other hand, the idealist purpose is "to have a real dialogue with the public about a future that remains genuinely open" (Jones 2007, p. 262). Hence, engagement idealists seek to ameliorate what they see as flawed methods of decision-making and opinion formation not only in political institutions but also among a deliberatively undisciplined public. Interestingly, neither of these purposes emphasizes understanding existing publics, except as such understanding mediates between limited-scope events and broader cynical or idealist purposes.

In both cases publics are prefigured as flawed; they must either be led to the right conclusion (cynical) or be taught the right way to deliberate (idealist). Torn between these competing political motivations for public engagement work, we lose sight of the role public engagement can play in understanding existing publics and how they come to opinion and action. This may be part of the reason why policy makers and funding agencies, as

© The Author(s) 2018
A.S. Lerner, P.J. Gehrke, *Organic Public Engagement*,
https://doi.org/10.1007/978-3-319-64397-7_4

Jones (2007) put it, "have been heard to remark that public engagement is a very expensive way of producing far from earth-shattering conclusions" (p. 263). While Jack Stilgoe (2007) is right that a growing interest in engagement spurred "the emergence of consultants eager to deliver democracy in neat packages," one wonders what democracy is being delivered and what public is being engaged (p. 75). These "experts of community voice" represent publics via standardized methods, such as consensus conferences, at the expense of methods that promise higher ecological validity and an understanding of how and why actual publics engage with issues (Felt and Fochler 2008, pp. 491–492).

While public engagement scholars have long expressed concern with degrees of scientific expert control in public engagement events, they rarely reflect that concern back upon the control of the public engagement experts themselves. Kleinman (2000) noted that approaches to studying public engagement with science vary in the degree of scientific expert control of the engagement, ranging from scenarios in which scientists exclude nontechnical issues, discount participants' inputs, and completely define the agenda of the event to grassroots social movements that inserted themselves into discussions of science and science policy (pp. 143–148). While this spectrum has become foundational to public engagement with science, the current crisis emerges from a growing body of public engagement experts who govern the conditions of engagement, a corollary problem of control.

This chapter investigates the theoretical and philosophical underpinnings of the idealist and cynical purposes of public engagement. Scholars in the field of public engagement with science have made note of the influence these two camps have made on our modern conception of public engagement. As Brian Wynne (2008) has suggested, public engagement with science scholars should "draw more fully on wider and more historical work in political philosophy, not only but including Dewey's response to Lippmann's dismissal of the public as merely a 'phantom'" (p. 21). Alternatively, Darrin Durant (2011) has argued one consequence of ignoring these political theories has been "the prevalence of thin conceptions of what constitutes a democratic sensibility in claims... about the relations between experts and publics in liberal democracies" (p. 692). In order to respond to Wynne's call for more investigation of the Dewey/Lippmann debate and answer Durant's call for increased attention to political theory, starting with Kant's conception of the public/private split, we examine two major debates in democratic theory: Walter Lippmann and John

Dewey's positions toward the public and John Rawls and Jürgen Habermas's dispute about the features of an ideal democratic form of engagement. Lippmann and Dewey's divergent views display one of the originating divides between cynical and idealist purposes that we now see in public engagement practices. Rawls and Habermas's dispute, however, represents two different takes on idealist modes of public engagement.

Having laid out the basic contours of these debates, we then move past the deficit model of public understanding of science. We turn to Amy Gutmann's work on deliberative democracy as a compelling philosophy that synthesizes these lines of influence and allows us to question the distinction between normative and descriptive engagement projects. We then contrast the philosophies of dominant practices in public engagement with science with the work of Michael Warner and Robert Asen on actually existing publics. In public engagement with science, scholars rarely attend to the implications of their version of what a "public" is and how it relates to theoretical and philosophical discussions. Few public engagement with science scholars have noted at length how the "public" in "public understanding of science" functions, or the constitutive boundaries that define the role of the public in relation to scientific experts. For instance, the deficit model posits a public and a set of experts and their constitutive epistemic positions with regard to science. However, what constitutes a public or an expert and how do we determine their epistemic position? Furthermore, how do we measure the gap (if any) between the two, and how does measuring this gap affect our understanding of both publics and experts? As Durant (2008) has argued, much of the public understanding of science literature has attempted to deny the gaps between experts and publics that are often presupposed in deficit models (p. 5). However, while much has been done to close the epistemic gap, in doing so, public engagement with science scholars now fly "close to an ontological gap," where the "lay public is modeled as reflexive agents whereas the expert is modeled as unreflexive" (Durant 2008, p. 5). Warner and Asen shed fresh light on these distinctions, close this ontological gap, and help tease out the often contradictory and paradoxical boundaries between publics and experts. Their shift toward discursive modes of understanding is a first step in developing a more ecologically valid mode of public engagement.

In order to further elaborate on Warner and Asen's potential contributions to the field of public engagement with science, we deploy the work of rhetoric scholars Samuel McCormick, Kent Ono, John Sloop,

Gerald Hauser, and Robert Glenn Howard. Rhetoric studies are relatively unrepresented in public engagement with science; however, there are some features of this field that lend itself to engaging in a discussion about the ecological validity of public engagement methods. In particular, rhetoric scholars' research on vernacular is an important articulation of ecologically valid data. While public engagement with science scholars has been clear to describe the difference between expert and public levels of understanding scientific topics, they rarely characterize the kinds of discourse these groups produce.

One of the important moves we make in this chapter is toward discursive and performative types of data. Much research in public engagement with science is grounded in epistemic concerns, like how much the public knows about a scientific topic, what their attitudes toward these topics are, or more generally their understanding of scientific topics. However, since public *understanding* of science has largely been supplanted by public *engagement* with science, this opens opportunities to better understand how publics encounter, use, perform, and communicate with and about science instead of simply what they think about science. In turn, the kinds of interactions that are valuable in constructing ecologically valid engagement events depends upon the kinds of discourses that publics generate and use when they interact with science. As opposed to "expert" discourses of science, we borrow the term "vernacular" to describe this kind of discourse.

This chapter begins to confront these challenges by outlining key assumptions in the philosophies of deliberative democracy that inform public engagement with science. We then turn to studies of publics and vernacular rhetoric from recent scholars in rhetoric studies. Together these constitute a compelling case for a method of public engagement with science that engages actually existing publics as they communicate and deliberate "in the wild."

PHILOSOPHIES OF DELIBERATIVE DEMOCRACY

Kant's Public and Private

Kant's 1784 "An Answer to the Question: What is Enlightenment?" is a short but highly influential text in political philosophy. The essay opens with a statement that cuts to the core of both idealist and cynical approaches to public engagement: "Enlightenment is the human being's emancipation from its self-incurred immaturity. Immaturity is the inability to make use of one's intellect without the direction of another" (Kant 2006, p. 17). Despite its brevity, these two statements are rich with meaning. Importantly, the term "immaturity" is a translation of *Unmündligkeit*, which can mean both a legal and inherent immaturity; *Mund* means "mouth," and a major connotation of *Unmündligkeit* is an inability to speak or decide for oneself. Human kind, therefore, suffers from a self-imposed incapacity to speak and make use of their intellect without the help of others.

In many ways, the first sentence shows why deliberative engagement methods are an important and valuable trend in public engagement with science. Starting from the premise that humans lack the ability to emancipate themselves from a condition of speechlessness, deliberative engagement methods promise that each participant will be able to utilize their intellect and weigh in on important issues in science, technology, and medicine. However, the second sentence, immaturity exists insofar as humans are not directed by others, demonstrates the beating heart of public engagement idealism and cynicism. On the one hand, Kant seems to find a kernel of democratic potential in the otherwise bleak situatedness of the human intellect; his famous claim "*Sapere aude!*" or "Have the courage to use one's own intellect!" could be the hypothetical tagline for any deliberative engagement. On the other hand, Kant seems to position the Enlightenment (as a historical epoch) as a relief from persistent human incapacity. Much like the cynical positions of scholars who engage with the public after decisions have already been made, Kant seems to suggest that the human reasoning has been bound by an unhealthy inability to demonstrate and express the intellect.

As Kant goes on to explain, when intellect is used *publicly* it should be unfettered by the bonds of regulations and norms. Alternatively, when it is used *privately*, reason is highly restricted by conventions and procedures dictated by organizations and entities to which an individual is bound by employment or participation. While this may seem like a strange breakdown of public and private (indeed, it is often remarked that they seem reversed), when we consider that Kant's conception centers on freedom, this distinction holds. For public engagement with science scholars who believe deliberative engagements are a method of alleviating the lack of public participation in scientific matters, this is important. As G.L. Ercolini (2016) notes, for Kant the public use of reason, or to speak publicly means speaking freely as a thinking and courageous person who submits arguments and ideas "in an accessible form for consideration and engagement" (p. 3).

In contrast, Kant's category of private speech is a position of obedience, where one speaks in a restricted role, such as a soldier, civil servant, or subject of a government. When one speaks from this position, Kant argued that reason and speech "may often be very narrowly restricted without particularly hindering the progress of enlightenment" (Ercolini 2016, p. 3).

In Kant's terms, by constructing elaborate methods of controlling, monitoring, or otherwise interfering with participant's actual and already existing ecologies, and setting participants into the role of a community representative or civil servant, public engagement practitioners are not so much doing *public* engagement, but instead subtly slipping into *private* engagement. According to Kant, public use of reason is one that is emancipated from the kinds of roles, duties, and responsibilities that one would normally find in the backdrop of these highly curated events. As Kant explains, a private citizen is required to pay taxes, but as a member of the public, "the same citizen does not contravene his civic duty if he publicly expresses, as a scholar, his thoughts against the impropriety or even injustice of such levies" (Kant 2006, p. 19). The more that participants' roles and modes of engagement are circumscribed, communicative norms are imposed, and reason occurs in a stilted and artificial environment, the more we move from the freedom of publicness to the regulations of the private. Kant, in other words, begins to demonstrate how despite the best intentions of public engagement practitioners, public engagement events seem to unfold in "un-public" ways.

Lippmann and Dewey

Kant provides a solid foundation from which to better understand the split between cynical and idealist purposes, because he seemingly spans both positions simultaneously. On the one hand, he seems to understand how the public is burdened by an intellectual and communicative incapacity (cynical), yet he also dares the public to transcend these limitations and utilize their voice in public matters (idealist). This contrast is clearly demarcated in the work of Walter Lippmann and John Dewey, prolific and well-known scholars of the early twentieth century. The best representation of their respective positions occurs in their works, *The Phantom Public* and *The Public and its Problems*, which were both published in the 1920s. Perhaps part of the impetus for their respective arguments can be drawn from the rapid shift in modes of communication and public participation occurring at the time. In 1927 alone, the first transatlantic telephone call is made from New York to London, the Federal Radio Commission began to regulate radio frequencies, the first live demonstration of television is made at the Bell Telephone Building in New York, and the end of the silent film era was signaled by the successful opening of *The Jazz Singer*. If Marshal McLuhan's basic premise that the "medium is the message" holds any credence, the late 1920s in America sends a message of rapid and total transformation of information, discourse, and publics.

This is the backdrop for Lippmann's *The Phantom Public*. Here, Lippmann develops a position of deep cynicism and skepticism toward the public's capacity to make decisions. His cynicism is unequivocal: "The individual man does not have opinions on all public affairs. He does not know how to direct public affairs. He does not know what is happening, why it is happening, what ought to happen. I cannot imagine how he could know, and there is not the least reason for thinking… that the compounding of individual ignorances in masses of people can produce continuous directing force in public affairs" (Lippmann 1927, p. 29). Much of this cynicism derives from Lippmann's perception of an increasingly complex and technical world. As Lippmann (1927) argues, democratic theorists unrealistically conceived of ideal citizens who are "expected to yield an unlimited quantity of public spirit, interest, curiosity and effort" (p. 14). They miss a critical fact: "the citizen gives but a little of his time to public affairs, has but a casual interest in facts and but a poor appetite for theory" (Lippmann 1927, pp. 14–15). Without a "rational ground for fixing his attention where it will do the most good, and in a way that suits his inherently amateurish equipment, he will be as bewildered as a puppy trying to lick three bones at once" (Lippmann 1927, p. 15). Hence, the framework for a cynical public engagement: have an expert tell the puppy which bone to lick. In a world becoming increasingly complex, Lippmann posits a "phantom" public, a public both ineffectual, incapable, and therefore necessitating the use of experts in policy decisions.

Dewey (1927) notes Lippmann's spectral metaphor in his pseudo-response, *The Public and its Problems*: "the public and its organization for political ends is not only a ghost, but a ghost which walks and talks, and obscures, confuses and misleads governmental action in a disastrous way" (p. 125). Dewey's counter-claim is to suggest that while "pertinent…to administrative activities," conceiving of the public as phantom-esque tends to "ignore forces which have to be composed and resolved before technical and specialized action can come into play" (Dewey 1927, p. 125). Instead of succumbing to Lippmann's cynicism, Dewey (1927) instead takes it as an opportunity to lay out the improvements needed to alleviate the problem of a phantom public, or the enhancement of the "methods and conditions of debate, discussion and persuasion" (p. 208). He concedes to Lippmann that this improvement requires experts, but clarifies that what these experts provide is not "shown in framing and executing policies, but in discovering and making known the facts upon which the former depend" (Dewey 1927, p. 209). In other words, the average citizen does not need to understand and contemplate the details of each public concern. Instead, they need only to "have the ability to judge of the bearing of knowledge supplied by others upon common concerns," the difficulty of which Dewey (1927) believes is overstated (p. 209). Experts provide the pertinent facts and frameworks for the public to deliberate, a skill that precedes the requirement of any technical know-how. Identifying a deliberative core that precedes specific epistemological demands, Dewey's perspective fits neatly into the category of idealist purposes of public engagement.

Those acquainted with public engagement practices should find these two perspectives familiar. Cynical purposes are implicit in engagement designs that produce no tangible outcomes, are designed to function "downstream" of the decision process, or otherwise attempt to eliminate negative views held by the public. For example, public engagement events focused on genetically modified organisms that actively dismiss or do not listen to public fears about this topic tend to be cynical in their purposes. Channeling Lippmann, instead of considering public views as legitimate, these public engagement events tacitly defer to expert opinions. On the other hand, idealist purposes are implicit in engagement designs that seek to "guide" the public toward constructive debate, argumentation, and deliberation. For instance, the inclusion of extensively curated introductory materials on a subject, the regulation of speaking times and formats, or any kind of restrictive parameters used to "direct" participants to deliberate in a particular way exhibit idealist purposes. Public engagement idealists view the public's opinions as legitimate, but only insofar as their opinions are processed through an ideal deliberative apparatus. Echoing Dewey, these public engagement events assume that the public has the capacity to make valuable judgments, but only if the information they are judging is curated by experts.

Both positions can be seen in contemporary engagement practices, and neither support strong ecological validity. Cynical purposes have fallen out of vogue in public engagement circles, but we could point to specific examples in the previous chapter that demonstrate the cynic's characteristic ineffectuality. Deliberative engagements need to trust that the public has something to contribute, believe that they are capable of making that contribution, and tangibly include their deliberations in the decision-making process. By not doing so, the public being engaged is spectral at best. Alternatively, idealist purposes abound in contemporary public engagement research. While it is heartening that these public engagement specialists respect the power of democratic deliberation, their attempts to fit the public to an ideal deliberative process weakens the ecological validity of their findings. To achieve even a moderate level of ecological validity, the deliberative process must be tailored for the specific public, not the other way around. By shaping the public to fit an idealist mold, these engagement events distort how deliberation actually occurs in publics' constitutive ecologies, altering the outcome of the deliberative process and skewing the data that emerges.

Rawls Versus Habermas

Just as Dewey once wrote that all philosophy must now respond to Kant, political theory in the twentieth century found itself obligated to respond to both Kant and Dewey. Deliberative democracy theorists, rhetoric scholars, and public engagement researchers have all found much inspiration from Kant and Dewey. John Rawls and Jürgen Habermas have been especially successful at building upon these thinkers, shaping Western political philosophy in the latter half of the twentieth century and laying the foundations of public engagement theory. As Durant has argued, most of the mainstream public engagement models are wedded to the philosophies of Rawls, Habermas, or in some cases both.

Much like Kant's categorical imperative, Rawls's *A Theory of Justice* marks an important attempt to devise a system of political participation grounded in rational morality. In *Political Liberalism,* he updates and elaborates on the basic theory he put forth in *Justice.* Here, he presents his famous "original position," a deliberative model that begins with the premise that in order to achieve justice (as fairness), some sort of ideal system of social cooperation must be theorized. He defines social cooperation as having three elements: First, cooperation is not ordered by a central authority, but instead guided by "publicly recognized rules and procedures those cooperating accept and regard as properly regulating their conduct" (Rawls 1993, p. 16). Second, cooperation requires a set of fair terms "that each participant may reasonably accept, provided that everyone else likewise accepts them" (Rawls 1993, p. 16). Finally, cooperation requires that each participant has their "rational advantage," or what each individual hopes to achieve by cooperating (Rawls 1993, p. 16). Rawls argues that the only way to achieve this sort of deliberative ideal is to invent "some point of view, removed from and not distorted by the particular features and circumstances of the all-encompassing background frame-work, from which fair agreement between persons regarded as free and equal can be reached" (Rawls 1993, p. 23). This "original position" is what he calls "the veil of ignorance," where participants are not allowed to know the social position, race or ethnic group, sex and gender, or any "native endowments" of anyone they are representing (including themselves) (Rawls 1993, p. 25). This produces a condition where participants' deliberations become unfettered by egoism, since they will make decisions on the basis of what is good to the most precarious and victimized classes of people, since they are "ignorant" of their own identity and therefore could very well be a member of a maltreated group. By establishing these conditions, Rawls hopes to provide a fair and equitable position from which participants in a deliberative situation can make ethical decisions.

Despite their relatively parallel intellectual trajectories, Habermas only responded to Rawls's democratic theories in an article published in 1995, late in both scholars' storied careers. Here, Habermas painstakingly addresses the minutiae of the original position, and problematizes Rawls's theory in two major ways. First, Habermas (1995) flatly states that he "cannot consistently stand by the decision that 'fully' autonomous citizens are to be represented by parties who lack this autonomy" (p. 112). According to Habermas, Rawls assumes that citizens have an innate capacity to be moral persons who possess a sense of justice and a desire for their own conception of the good, but by entering the original position, these citizens are stripped of such features. Therefore, a citizen stripped of these qualities cannot accurately represent a group of citizens who still retain these essential characteristics. By equalizing the playing field via the original position, Rawls has unintentionally short-circuited the capacity for accurate representation. Second, Habermas (1995) argues that Rawls "imposes a common perspective on parties in the original position through informational constraints and thereby neutralizes the multiplicity of particular interpretive perspectives from the outset" (p. 117). This is a more straightforward critique and one where Habermas demonstrates his theoretical alliances. Arguing for a theory of discourse ethics, Habermas (1995) views "the moral point of view as embodied in an intersubjective practice of argumentation which enjoins those involved to an idealized *enlargement* of their interpretive perspectives" (p. 117). For Habermas, "flattening out" the deliberative process into an ideal procedural mechanism ends up defeating the very purpose from which this procedure is created. Habermas's (1995) alternative to Rawls's idealism is to establish conditions whereby "everyone is required to take the perspective of everyone else, and thus project herself into the understandings of self and world of all others" (p. 117). These conditions appear in discursive exchanges rather than deliberative and political decision-making practices.

Both Rawls and Habermas express idealistic forms of public engagement, but frame their ideals in different terms. For Rawls, establishing a procedural and systematic form of equalization prior to deliberation is key for producing ethical and valuable outcomes. For Habermas, producing this form of systematic leveling is counter-productive; instead, the idealist paths into which deliberative participants are compelled emerge through the intersubjective collusion of participants in ethical debate and discourse. In either case, these idealist modes of conceiving of deliberation are salient in many engagement practices. Attempts to produce equalized deliberative footing for all involved, not only as individual participants but as representatives for their communities, utilizes a steadfastly Rawlsian approach. By priming deliberative procedures to make engagement events more "fair," engagement practitioners are implicitly engaged with a Rawlsian political theory. Alternatively, regulation of discursive norms, such as setting time limits on speaking or encouraging certain types of discourse over others, is tapping into a classic Habermasian view of discourse ethics. Many engagement events utilize both idealist frameworks simultaneously.

The idealist frameworks that these political theorists put forth, while noble in their attempts to produce conditions of fair, equitable, and just deliberation, weaken the ecological validity of their results. As Chantal Mouffe (2009) has argued, in Rawls's form of liberalism conflicts of interest about economics or social issues are "resolved smoothly through discussions within the framework of public reason, by invoking the principles of justice that everybody endorses" (p. 7). However, if "any unreasonable or irrational person happens to disagree with that state of affairs and intends to disrupt the consensus, she must be forced, through coercion, to submit to principles of justice" (Mouffe 2009, p 7). Since the participant is being "unreasonable," Mouffe (2009) argues that under Rawls's conception "this type of coercion...does not entail oppression" (p. 7).

This point is illustrative for public engagement events. If the goal of these events is to create a space where democratic deliberation occurs, just how democratic are these spaces if they do not allow unreasonable or irrational critiques of their procedures? Participants have two options if they disagree: either they are coerced into following the procedures put forth by organizers, or, they recuse themselves from the proceedings entirely. In either case, the resultant data that is derived from these events will necessarily be skewed, because in reality, varied people and publics deploy divergent rationalities and standards for reasonableness, and these influence the course and direction of debate. Regardless of how meticulously public engagement organizers may try to capture the demographics they seek to represent, the "emotional," "irrational," or "unreasonable" participants will always be put into a position of coercion. As Mouffe (2009) suggests, "such a scenario presupposes that political actors are only driven by what they see as their rational self-advantage" (p. 8). Such an assumption severely undermines the ecological validity of public engagement research.

On the other hand, Habermasian ideals fare no better according to Mouffe. A Habermasian discourse model states that for deliberative arrangements to be valid, they should fulfill three basic criteria: first, all must have the chance to engage in speech acts. Second, all have the right to question the assigned topics of discussion. Third, all have the right to question the rules and discursive procedures. If these criteria are fulfilled, Mouffe (1999) suggests these "values will guide the discussion towards generalizable interests to the agreement of all participants…[which] can be guaranteed to have reasonable outcomes only to the extent that it realizes the conditions of ideal discourse" (p. 748). As opposed to Rawls, Habermas's model seems to have the upper hand with regard to ecological validity, since participants in a deliberative situation will have the capacity to rewrite the discursive rules under which this deliberation occurs. However, Mouffe suggests two lines of argument that put the potential ecological validity of Habermas's model in question. The first is derived from Wittgenstein's concept of language games.

As Mouffe explains, Wittgenstein's view of agreement does not rely on significations and rationality, but instead on a "form of life" or *Lebensform*. Agreement in a Wittgensteinian model is centered on *Stimmung*, "mood," more than it is on *Verstand*, "understanding." This approach would require "reintroducing into the process of deliberation the whole rhetorical dimension that the Habermasian perspective is precisely at pains to eliminate" (Mouffe 1999, p. 749). Without this dimension, the Habermasian model stands face-to-face with the very limits of consensus: how does the Habermasian model cope with irreconcilable differences? How can it deal with inevitable result of this fundamental conflict, where participants begin to question the very rationality of their opponents? Habermas and Rawls have no way of democratically or rationally dealing with these types of encounters. Divergent rationalities are a part of our everyday ecologies, yet neither of these models is capable of folding "irrational" thought into their overwhelmingly rational systems. A second line of argument Mouffe directs toward Habermas is derived from the psychoanalytic theories of Jacques Lacan.

According to Mouffe (1999), a "Lacanian approach reveals how discourse itself in its fundamental structure is authoritarian since out of the free-floating dispersion of signifiers, it is only through the intervention of a master signifier that a consistent field of meaning can emerge" (p. 751). Habermas and Rawls assert the democratic and egalitarian nature of their models, but both are always already authoritarian since they "imply the idea of a communication free of constraint where only rational argumentation counts" (Mouffe 1999, p. 751). Describing the norms of deliberative engagement or forcing participants to adhere to an ideal of engagement both constitute an "intervention of a master signifier," which in turn disrupts the very democratic fabric that these interventions are meant to produce. In either case, both the Habermasian and Rawlsian models of deliberation fall short in terms of their ecological validity, since varied publics use divergent rationalities, participants are not always amenable, and attempting to modifying either of these conditions drives a wedge between lived reality and the public engagement event.

How then can deliberative engagement cope with these fundamental problems? Amy Gutmann's and Dennis Thompson's (1998) work, which deals with the question of deliberative disagreement, suggests that when participants deliberate with one another, they should seek moral agreement when possible and maintain mutual respect when they cannot (p. 346). The goal of deliberative engagement should not be to eliminate or alter the conditions of conflict like the idealist models of Habermas and Rawls. Rather, deliberative engagement must take account of how conflict occurs and how discussion can and does fall into disrepair. This disrepair, importantly, is not a *failure* but a *feature* of ecologically valid engagement events. Public engagement cynics interpret this failure as a justification for ignoring or discounting participant's views, while idealists go to great lengths to reduce the possibility of this failure occurring in the first place. In either case, there is a reluctance to account for the actually existing ecologies of everyday life. Part of the failure of these methods is their inability to account for how actually existing publics function. Instead of accounting for how a public functions and creating a method to fit the contours of a particular public's functionality, these methods instead seek to create a public out of ideals. As Mike Michael (2009) has rightly argued, publics are created "as particular types of citizen by virtue of the models of the public that inform public engagement with science initiatives" (p. 619). This inhibits our capacity to engage organically and ecologically with publics. In order to set a groundwork for calibrating methods to publics, we should begin by investigating what we mean by a "public."

PUBLICS, COUNTERPUBLICS, AND VERNACULAR RHETORIC

Publics and Counterpublics

Michael's (2009) insightful differentiation between what he calls Publics-in-Particular (PiPs) and Public-in-General (PiGs) is one of the more sophisticated attempts to outline the basic feature of publics in public engagement with science scholarship. To put his analysis simplistically, PiGs are constituted in relation to science "in general," while PiPs are constituted in relation to a particular scientific debate, question, or enterprise. The deficit model, for instance, taps into the idea of a PiG, because the "public" in question is amorphous, or what a semiotician would call a "floating signifier." On the other hand, PiPs are "demarcated in relation to some external event" (Michael 2009, p. 623). The residents in and around Chernobyl, for instance, could be classed as a PiP because of their direct relationship with a spatial or geographical technoscientific problem or concern. As Michael (2009) cautions, however, PiPs are not "transparently obvious entities" (p. 625). Citing Wynne, Michael explains how some PiPs' identities can be conflated, for example by attributing incompetence to a group who used toxic chemicals rather than critiquing the toxicity of the chemicals themselves. Michael's characterization of PiGs and PiPs leaves the term "publics" elusive, but we can get a better sense of this distinction by drawing on more robust and theoretical treatments of publics.

Michael's breakdown of publics as PiPs and PiGs is in no small way influenced by the work of Michael Warner. As Warner suggests, there is a difference between *a* public (PiP) and *the* public (PiG). As Warner (2005) proposes, *the* public "is a kind of social totality… [which] is thought to include everyone within the field in question" (p. 65). This can include everything from a nation, to a commonwealth, to a city, state, or region. On the other hand, *a* public is "a concrete audience, a crowd witnessing itself in visible space…bounded by the event or by the shared physical space" (Warner 2005, p. 66). Much like Michael, Warner conceives of this public as occupying shared geospatial commonalities. However, Warner's work is concerned with a third sense of public: the kind that is generated in relation to texts and their circulation. While Warner (2005) admits that the distinction between "these three senses are not always sharp" (p. 66), by positioning the basis of publics on the discursive, it opens a descriptive resource that can help shape our understanding of what a public is, and help us determine if public engagement with science is indeed about engaging with publics as they actually exist. Warner sets out seven basic criteria for what constitutes a public, but there are three main characteristics that are relevant for setting up a groundwork for public engagement with science scholars.

First, a public is self-organized, or, "it exists *by virtue of being addressed*" (Warner 2005, p. 67). Warner (2005) immediately addresses the paradoxical circularity of this assertion: "How can the existence of a public depend, from one point of view, on the rhetorical address and, from another point of view, on the real context of reception?" (p. 67). Warner sees this not as a problem, but as a feature of how a public functions. A public is constituted by a reflexivity not established by any institution, law, formal citizenship, or organization. In short, a public is a "space of discourse organized by discourse… It is self-creating and self-organized; and herein lies its power, as well as its elusive strangeness" (Warner 2005, pp. 68–69). Warner notes how pollsters and social scientists have long attempted to define and study publics empirically, independently of the reflexive discursivity by which these publics define themselves. However, this has turned out to be harrowingly difficult, since a practical infinite of publics can exist and people can belong to different publics simultaneously. A public is not easily constrained by physical space, since, by definition, a public is understood to be different from an entity that requires co-presence, like an audience or crowd. On the other hand, personal identity does not necessarily make oneself part of a particular public, since publics are different from nations, professions, or races. Instead, Warner claims that belonging to a public simply requires minimal participation, however notional. Even arguing that a public is definable in terms of common interests is problematic, because it "appears to identify the social base of public discourse; but the base is in fact projected from the public discourse itself rather than external to it" (Warner 2005, p. 71). Despite its seeming vagueness, Warner's first characterization of a public sets a tentative boundary for how publics function.

Second, Warner (2005) suggests that a public is "*the social space created by the reflexive circulation of discourse*" (p. 90). More than a simple model of sender/receiver or author/reader, Warner's (2005) conception of a public goes "beyond the scale of conversation or discussion to encompass a multigeneric lifeworld organized not just by a relational axis of utterance and response but by potentially infinite axes of citation and characterization" (p. 91). In addition to the complex flows of discursivity that define publics, Warner notes that part of this flow requires a mode of circulation. This performative ability, in turn, is what gives publics their characteristic paradoxical form, where they are both creating and being created by the channels of discourse that surround them, are founded within them, and intercede on them from every external angle. In turn, this circulation is bound by temporality: discourse moves in fits and starts, pauses and rhythms. Publics are shaped by a nomadic discursivity and citationality. While not entirely distinct from the first aspect discussed here of Warner's public, this notion of motion and circulation also posits a concrete (yet permeable) boundary that can help us better pin down a rough sketch of a public's major functions.

Third, "*a public is a poetic world making*" (Warner 2005, p. 114). A public is constituted through the "pragmatics of its speech genres, idioms, stylistic markers, address, temporality, *mise-en-scène*, citational field, interlocutory protocols, lexicon, and so on" (Warner 2005, p. 114). This performative dimension of publics exists outside of the kind of rational dialogic formulation that is indicative of thinkers like Rawls and Habermas. In rational paradigms, publics "exist to deliberate and then decide," however this obscures the essential poetical functions of publics, in both their language and expressivity (Warner 2005, p. 115). As Warner suggests, this particular language ideology favors sense, and acts of reading or listening are considered uniform, as are opinions. This normative vision of a public lends itself to a peculiar political situation, whereby publics that most closely follow the dictums of a rational conception of discourse become representative of *the* public. This is largely understandable: in an ecology marked by discursive and epistemic diversity, there is a pragmatic upside to constructing a rational ideal of public discourse and propping up certain publics as being more or less representative of what *the* public is. The problem with this, of course, is the conflation between descriptive and normative modes of understanding publics. The infinite diversity of potential publics makes finding representative publics a perpetually normative exercise, which in turn eliminates the potential to generate a picture of publics based on representative models.

Through their use of idealist models of public engagement, public engagement practitioners have eliminated, constrained, or otherwise stifled the very qualities that define a public. In the first case, the strict containment and guidance of participants in engagement events by public engagement practitioners limits the capacity for publics to engage reflexively. By constructing these limitations, public engagement events prevent the artificial aggregation of representative members of *the* public from ever truly forming into *a* public. That is, the members of a public engagement event do not represent *the* public ("the" public cannot exist as anything more than a projection of an ideal), but they are also prevented from developing into *a* public, because of the discursive limitations placed on these events. Participants are handed the information they "need to know." Discussion ("dialogue") is manufactured under the auspices of an artificial rational debate. As *a* public, this aggregation of demographics is placed in a perpetual state of stasis, incapable of reflexively defining the boundaries of their discursive constitution. Deeply related to this problem, by constraining this potential public to the ideals of democratic participation in a "rational" system, the circulation of discourse is stymied. The complex flows of discourse that we might observe emerging and influencing a public "in the wild" are very different from that we see in a consensus conference. When participants are constrained to such ideals, the wide variety of unexpected and (perhaps) unwanted or uninitiated twists and turns of actual discussion are nearly eliminated or simply ignored. Finally, while participants in public engagement events surely perform the kinds of pragmatic and poetic characteristics of a public that Warner describes, these features do not fit neatly into a rational/dialogic paradigm and are therefore considered "white noise" or "extraneous." Few if any researchers who investigate public engagement events will remark on the kinds of performative events that help define the "mood" the event took on. Instead, they often fit discursive themes into typologies, or cherry-pick phrases, words, or ideas that fit their idealist understanding of rational discourse. It is perhaps ironic that such earnest attempts to engage with a public frequently end up obscuring those features that define and characterize a public in the first place.

Public engagement practitioners have long attempted to ameliorate what many theorists have argued is a declining public involvement in communities and deliberative processes relevant to citizens. In public engagement with science, this is certainly the case: many efforts to bring the public into the fold on scientific topics are an attempt to "close the gap" between scientists and laypeople. These efforts make assumptions about what counts and does not count as "proper" engagement. Hence, starting from the premise that publics and scientific experts are not engaging in a productive dialogue may lead to idealist purposes for public engagement, since this premise already assumes that the *type* of engagement is insufficient, unsuitable, or otherwise lacking. In other words, these public engagement events often tacitly suggest that certain kinds of engagement with science count, while others do not. As Robert Asen (2005) argues, however, asking the "question of what counts as citizenship does not admit degrees of difference, yet the same activity may express a different meaning and significance in various contexts" (p. 191). At the heart of both cynical and idealist public engagement purposes is the erasure of engagement contexts. The *prima facie* notion that engagement can change depending on the ecological conditions is ostensibly eliminated.

Much like Warner, who grounds his theory of publics in discursivity, Asen proposes a discourse theory of citizenship. Asen (2005) argues that rather than "asking what counts as citizenship, we should ask: how do people enact citizenship?" (p. 191). In making this shift from counting to functioning, we shift from a prescriptive to a descriptive framework. As Asen (2005) suggests, prescribing forms of participation risks denying subjectivities that lie outside of dominant norms (p. 192). Description too has the potential to be erroneously universalized, however. Importantly, subjectivities are context-dependent. By focusing on the discursive features of subjectivities, this becomes less of a problem, since we no longer emphasize demographic qualities or features "innate" to certain groups of people, but instead hone in on how subjectivities are actively formed by reflexive, circulatory, and world-making discourse. Asen suggests that citizenship is itself a *mode* of public engagement. Modes are already always multiple; to become a public subject is to engage through "work habits, consumption patterns, and familiar interactions" (Asen 2005, p. 195). As he eloquently states it, "insofar as subjectivity emerges through our interactions with others, and insofar as our interactions with others are varied, and insofar as we behave differently in these different interactions, we create different public selves" (Asen 2005, pp. 195–196). Public engagement practitioners go to great lengths to reduce the variety of modes in which citizens engage publicly with science, often by utilizing rarefied forms of public engagement founded on cynical or idealist purposes. As Asen might argue, these events are not really public engagement at all, since the central feature of public engagement in a discourse theory of citizenship is *risk*. Public engagement with science events usually seeks to reduce risk, eliminate risk, or divert risk into narrow channels where it can be managed and controlled. The riskiest kinds of discourse are those that do not fit into the neat boxes of rational deliberation. As we have argued above, the performative, "irrational," or "unreasonable" kinds of activities that constitute actually occurring deliberation and discussion are constrained for the sake of producing an idealist or cynical model of public engagement. Another term for this kind of everyday discourse (a term that expands our understanding of how discourse is actually utilized) is "vernacular."

Vernacular Rhetoric

An ecological mindset will inevitably be drawn to vernacular data. In many cases, the data that is abandoned or marginalized in experimental settings is none other than the "everyday" sorts of social interaction that permeate actually occurring situations. In Chap. 2, we explored how ecological methodologies empowered sociologists and psychologists to examine common occurrences and social interactions. In Chap. 3, we outlined several studies, especially the work of Irwin, Wynne, Hine, Anderson, Delborne, and Kleinman, that pay close attention to the "informal" or "everyday" interactions that appear in digital and deliberative spaces. In this chapter, we have critiqued the pseudo-deliberative models of public engagement of both cynical and idealist varieties, which impede our access to a robust and ecologically valid data set. What is at stake in each of these cases is the fundamental tension between rarefied and vernacular data. Including both modes of social interaction increases the strength of a study's ecological validity, while limiting, constraining, or paring either of these types of data risks reducing the capacity of a study to reflect actually occurring conditions of deliberation, discussion, and dialogue. Warner and Asen's contributions to our understanding of how publics function (which is to say, what publics are) entreats us to capture all aspects of these conditions if we wish to engage with actually existing publics. One method of fulfilling this goal is by accounting for vernacular in public engagement research.

The most commonsense articulation of vernacular is McCormick's (2003) definition of "everyday talk" or "the ordinary kinds of communication people do in schools, workplaces, and at public meetings, as well as when they are at home or with friends" (p. 109). As he explains, "everyday talk" is distinct from "official patterns of discourse," or the "well-known and widely circulated ways of speaking characteristic of any given sociocultural context" (McCormick 2003, p. 109). For McCormick (2003), vernacular is constituted by "paralinguistic markers such as hesitations, repetitions, repairs, intonation, and emphases," which function to express the "subtle, often fleeting displays of emotion and spur-of-the-moment decisions that riddle public speech" (p. 109). Most importantly, for McCormick these markers (often imitative in nature) can provide insight into how historical and cultural forces are invoked in dimensions of communication we would otherwise ignore. Given this characterization of vernacular, attending to and taking seriously the hesitations, inflections, pauses, and slips of everyday speech in engagement settings affords us an opportunity to examine how historical, cultural, or institutional discourse resonate in specific publics. In other words, the stifling artificiality of engagement events, coupled with a tendency to overlook subtle and everyday expressions of meaning, results in a stilted view of how publics deliberate in their native ecologies. Such circumstances result in research with weak ecological validity.

McCormick lays an excellent groundwork for characterizing vernacular, but we can expand on his definition by examining Ono and Sloop's "Critique of Vernacular Discourse," which establishes some of the initial boundaries of this mode of communication. In particular, they make sure to point out how vernacular "does not function solely as oppositional to dominant ideologies" (Ono and Sloop 1995, p. 22). As our above discussion of Rawls and Habermas suggests, oppositional discourse is certainly one mode of discussion that is often erased or constrained in public engagement settings. However, Warner and Asen's analysis implicates almost all public discourse, not just oppositional or marginalized communication, in the artificiality of public engagement events. Many subjectivities are in fact marginalized by this artificiality, because publics are not capable of discursively functioning as publics in most public engagement settings. The lack of ecological validity at these events creates a problematic stifling of the "syncretic (protesting and affirming) aspects of discourse," which is how Ono and Sloop (1995) characterize vernacular (p. 38). Expanding on McCormick's definition of vernacular, these syncretic features include modes of pastiche, or the creation of "unique discursive form[s] out of cultural fragments" (Ono and Sloop 1995, p. 23). In practice, accounting for Ono and Sloop's definition of vernacular means examining the often-overlooked use of imitation or parody, irrespective of their seeming super-fluousness. Furthermore, as Ono and Sloop (1995) discovered in their complex investigation of women's resistance in Japanese American culture, "a critique of vernacular discourse can begin to illustrate how construc-tions are not necessarily, because they are vernacular, liberatory" (p. 38). Therefore, including vernacular in our investigations of public engagement with science is not reducible to moments of resistance or protest.

Although Ono and Sloop (1995) emphasize that vernacular rhetoric studies have an obligation to study "discourses that resonate within and from historically oppressed communities," this emphasis was never intended to come at the expense of localized communities more generally (p. 20). Instead, echoing McCormick's basic claim, this emphasis serves to offset the tendency to ignore certain communities and particular discourses in studies of publics (Ono and Sloop 1995, p. 20). All publics are, in some sense, vernacular publics. Each has its own methods of speaking, of arguing, of valuing, and of making sense. As we have argued from the beginning of this book, public engagement with science can gain from accounting for eco-logical validity, and we do not wish to suggest that current methodologies are necessarily oppressive, tyrannical, or despotic. On the contrary, there is much about deliberative engagement models that are inspirationally opti-mistic about the power of citizens' involvement in science. Rather, we wish to suggest that while the intentions behind this research have been laudable, the way in which this research is performed often creates paradoxical prob-lems that skew data, eliminate actual engagement, and constrict a public's functionality so much so that it no longer functions as a public.

Concern for actual publics and vernacular discourses requires taking up the empirical observational task. While vernacular rhetoric research moves past the analysis of individuals and corresponding aggregation indicative of current engagement research, one valid criticism is its com-mon overemphasis or reliance on textual evidence, such as newsletters or public documents, rather than observations of publics and their dis courses as lived practices. If not carefully conducted the study of vernacu-lar rhetoric can become a form of textual criticism that, as Gerald Hauser

(1999) argued, commits an "intentional fallacy" by failing to go "beyond the critic's own sensibilities" (p. 276). Using observational methods (quantitative or qualitative) and taking an ethnographic caution to minimize disturbances in the ecology of that public, we can study, as Hauser (1999) put it, "how actual publics respond to appeals, how they themselves actually engage in discourse that allows us to infer their opinion, and the rhetorical conditions that color their interactions" (p. 20). Or, as McCormick might argue, we can better understand the kind of communication that occurs in everyday contexts.

McCormick's version of vernacular encourages us to contend with paralinguistic markers, while Ono and Sloop caution against assuming that vernacular is necessarily marginalized, and expand the definition of vernacular to include moments of pastiche. However, Hauser's view of vernacular has been characterized as "a diffuse set of communication practices generally available" (Howard 2010, p. 243). In particular, Hauser's *Vernacular Voices* articulates a theory of vernacular as a principle of invention, closely associated with the *doxa* or common sense of a public. He deploys this principle to point out a central difficulty in assessing the public's opinion via surveys or expert discourse, mainly, that they give "a distorted picture both of what counts as public opinion and of what we mean by the qualifier *public*" (Hauser 2007, p. 334). Much like we argued in Chap. 3, opinion polling for Hauser can be accurate in tracking views in a given timeframe as well as variance in attitudes held about a topic. However, Hauser also grasps that opinion polling falsely assumes that publics have opinions on the questions being asked, and that all participants' opinions carry equal weight. Moreover, surveys do not provide researchers with data on why people hold certain opinions about a subject, or the narratives that make their opinions matter to them. On the other hand, expert voices often change when leaders and workers (or for our purposes, experts and participants) discuss their opinions in separate spaces. For example, Hauser (2007) notes that elite voices with rhetorical acuity often have access to public media, attracting credibility and authority that they might otherwise lack if not for their bombastic and ratings-driven discourse (p. 335). In both the case of surveys and expert voices, the vernacular facets of a given ecology are not authentically represented.

Hauser (2007) expertly defines the place of vernacular in deliberative research:

> Complex public problems bring multiples perspectives to bear, each with its own understanding of what constitutes the salient issue. They invite an intersection of various interest publics, each with its own interests at stake.

Their deliberations occur in multiple discursive arenas spread across society and, in some cases, across national borders. Within these arenas, ordinary citizens engage in ongoing dialogue about the conditions that intersect with their lives and who are able to understand and respond to the *vernacular* exchanges that exists outside of power and are normative of it. (p. 335)

Hauser's articulation of vernacular neatly corresponds to the standards for engagement methods that promise stronger ecological validity. Vernacular modes of discourse are how publics engage in world-building in actually occurring situations, and current public engagement methodologies often overlook these quotidian data sets or limit the capacity for this data to emerge in the first place. Synthesizing McCormick, Ono and Sloop, and Hauser's articulation of vernacular, we can characterize this type of communication by broadly describing it as "noninstitutional" modes of expression. The internet is one space in which noninstitutional expression has flourished, especially given the rise of social media and web 2.0, both of which foster and provide a platform for institutional and noninstitutional discourse to have the same disseminative capacity.

As we argued in the last chapter, digital spaces of engagement present an increasingly important future direction for public engagement with science, and the internet offers public engagement experts a rich resource of vernacular discourse. As Howard (2010) explains, while websites sometimes offer institutional settings for discourse, the inclusion of collaborative features like comment sections offer a space in which vernacular can emerge (p. 245). Using the example of the official blog for the 2004 US presidential candidates John Kerry and John Edwards, Howard notes how these institutional-sponsored spaces blur the distinction between the vernacular and the institutional. On the one hand, participatory features like comment sections allow publics to express their opinions and engage in multiple dialogues, in turn framing the visual and semiotic ecology of a blog. On the other hand, blog owners can curate the comments sections of their websites, selecting which public's voices are heard or ignored. Alternatively, bloggers also have the opportunity to respond to comments. By engaging a public's discourse in the vernacular setting, institutions and institutionally backed messages can take on a vernacular character.

The hybridity or instability of the difference between vernacular (or "public") and institutional (or "expert") discourse is often highlighted in digital interactions. However, this hybridity, as Howard points out, has always been a part of the vernacular. The term "vernacular" has its root in the Latin term *verna*, which specifically referred to slaves that were born and reared in a Roman home. Sometimes these children were literally a cross between a Roman master and an enslaved servant, other times they were simply enslaved as children and took on the cultural characteristics of a Roman citizen. Often taught the institutional and cultural languages of Rome, these hybrid slaves were considered more valuable and powerful than their "outsider" counterparts. In fact, the Latin term *hibrida* is closely tied to *verna* (Howard 2010, p. 250). A *hibrida* can designate the cross between a domestic and "wild" animal, but in ancient Rome usually meant the cross between a sow and a wild boar. The boar was considered a masculine ideal in both Rome and Greece, and was thought to be a worthy and respected animal of the hunt. The term *hibrida* was sometimes applied to *verna*, who would have been perceived as partially tamed and partially wild. This hybridity was seen as powerful to some Romans, because these *verna* had access to both institutional and noninstitutional modes of communication. As Howard (2010) explains, because vernacular is always rendered distinct from the institutional, it can never technically be separated from the institutions it inheres in (p. 251). Howard differentiates his version of vernacular from Ono, Sloop, and Hauser by dubbing the mutual co-constituency of vernacular and institutional discourse as a "dialectical" vernacular.

Applied to public engagement with science, Howard's astute observations about the hybrid nature of vernacular puts the stark contrast between expert and lay forms of discourse in question. One way of working past the deficit model is to examine the ways that publics actually discuss scientific topics, which in turn can flip "expertise" on its head. Recall the short example of nuclear materials workers described by Wynne in the last chapter. While they did not know how to differentiate between kinds of radiation, they did know how to safely manage and handle dangerous materials. In fact, they argued that knowing this kind of "expert" information could potentially jeopardize their safety. In this scenario, who are the experts? The scientists that create the guidelines and safety procedures? Or the workers who deploy these procedures in practice? The assumption that "expert" discourse is somehow superior does not hold, nor does it hold that this particular public's actions are "nonexpert." A team of employees who follow precise safety guidelines seem as much an "expert" on handling nuclear materials as the scientists who generate those safety guidelines, regardless of how much they know about the science behind the rules.

In fact, the astute observation that this knowledge is a potential hazard demonstrates their expertise in dealing with dangerous nuclear materials. Expert and vernacular modes of understanding and communication largely take on these characteristics based not on their essential components but instead on the context in which they appear and emerge. An ecologically valid public engagement method simply seeks to capture the ecology at hand, vernacular and expert data included. Having collected this data, the ecologically valid engagement can *then* decide what is really "expert" and what is really "vernacular," based on actually occurring situations instead of prescribed notions of what counts as expert and what counts as vernacular.

Questions of what constitutes a public and how best to understand and engage them, as discussed in this chapter, do assume something like the idealist's view that publics are worthy of engagement and can make a meaningful contribution to process and policy. Even if idealist engagement theorists and practitioners believe the public needs to learn better methods of deliberation to produce valid contributions, they demonstrably believe that this is both possible and important to our civic life. Though efforts to democratize science and science policy may require adjustment to more ecologically valid methods in order to reach actually existing publics, they are at least ostensibly interested in those publics and value their potential contribution.

However, as we noted at the beginning of this chapter, not all public engagement with science holds this positive view of publics. Cynical public engagement starts with a deeper distrust of publics and their contributions, one that finds fault not so much with their methods of deliberation as with their status as nonexperts, as nonscientists. When technicians, scientists, or bureaucrats decide the course of policy in advance and then only engage publics as a public relations campaign or to check off a required part of the process, they reflect both a distrust of publics' capacities and also a belief in a divide between the experts (scientists) who know the best course of action and the nonexperts (publics) who must be fed the correct decision. This kind of "engagement" is so common in practice that both sides often complain about the uselessness of the engagement events. Returning from working with a local civic group conducting their own public engagement events on nanotechnologies, one of the authors met a retired engineer who worked for a large and generally progressive state. When the author told the engineer about the public engagement events, the engineer rolled his eyes, snorted, and said that he had hated doing those events because the state agency had always decided what was going to happen before the event was even scheduled, and people who attended did not know what they were talking about. He called them "a total waste of time." In some ways, overcoming this cynicism is more challenging, especially given the common published data on public knowledge and understanding of science. We confront this challenge in the next chapter through an examination of the demarcation problem: the problematic boundaries between the scientific and the nonscientific as well as between the expert and the nonexpert.

REFERENCES

Asen, Robert. 2005. Discourse Theory of Citizenship. *Quarterly Journal of Speech* 90 (2): 189–211.

Dewey, John. 1927. *The Public and Its Problems*. Chicago: Gateway Books.

Durant, Darrin. 2008. Accounting for Expertise: Wynne and the Autonomy of the Lay Public Actor. *Public Understanding of Science* 17 (1): 5–20.

———. 2011. Models of Democracy in Social Studies of Science. *Social Studies of Science* 41 (5): 691–714.

Ercolini, Gina L. 2016. *Kant's Philosophy of Communication*. Pittsburgh: Duquesne University Press.

Felt, Ulrike, and Maximillian Fochler. 2008. The Bottom-Up Meanings of the Concept of Public Participation in Science and Technology. *Science and Public Policy* 35 (7): 489–499.

Gutmann, Amy, and Dennis Thompson. 1998. *Democracy and Disagreement*. Cambridge: The Belknap Press of Harvard University Press.

Habermas, Jürgen. 1995. Reconciliation Through the Public Use of Reason: Remarks on John Rawls's Political Liberalism. *The Journal of Philosophy* 92 (3): 109–131.

Hauser, Gerald A. 1999. *Vernacular Voices: The Rhetoric of Publics and Public Spheres*. Columbia: University of South Carolina Press.

———. 2007. Vernacular Discourse and the Epistemic Dimension of Public Opinion. *Communication Theory* 17: 333–339.

Howard, Robert Glenn. 2010. The Vernacular Mode: Locating the Non-instituitional in the Practice of Citizenship. In *Public Modalities: Rhetoric, Culture, Media, and the Shape of Public Life*, ed. Daniel C. Brouwer, 240–261. Tuscaloosa: University of Alabama Press.

Jones, Richard. 2007. What Have We Learned from Public Engagement? *Nature Nanotechnology* 2 (5): 262–263.

Kant, Immanuel. 2006. An Answer to the Question: What Is Enlightenment? In *Toward Perpetual Peace and Other Writings on Politics, Peace, and History*, ed.

Pauline Kleingeld, trans. David L. Colclasure. New Haven: Yale University Press.

Kleinman, Daniel. 2000. Democratization of Science and Technology. In *Science, Technology, and Democracy*, ed. Daniel Kleinman, 139–165. Albany: State University of New York Press.

Lippmann, Walter. 1927. *The Phantom Public*. New Brunswick: The Macmillan Company.

McCormick, Samuel. 2003. Earning One's Inheritance: Rhetorical Criticism, Everyday Talk, and the Analysis of Public Discourse. *Quarterly Journal of Speech* 89 (2): 109–131.

Michael, Mike. 2009. Publics Performing Publics: Of PiGs, PiPs and Politics. *Public Understanding of Science* 18 (5): 617–631.

Mouffe, Chantal. 1999. Deliberative Democracy or Agonist Pluralism? *Social Research* 66 (3): 745–758.

———. 2009. The Limits of Jon Rawls' Pluralism. *Theoria: A Journal of Social and Political Theory* 56 (118): 1–14.

Ono, Kent A., and John M. Sloop. 1995. The Critique of Vernacular Discourse. *Communication Monographs* 62 (1): 19–46.

Pidgeon, Nick, and Tee Rogers-Hayden. 2007. Opening Up Nanotechnology Dialogue with the Publics: Risk Communication or 'Upstream Engagement'? *Health, Risk & Society* 9 (2): 191–201.

Rawls, John. 1993. *Political Liberalism*. New York: Columbia University Press.

Stilgoe, Jack. 2007. *Nanodialogues: Experiments in Public Engagement with Science*. London: Demos.

Warner, Michael. 2005. *Public and Counterpublics*. New York: Zone Books.

Wynne, Brian. 2008. Elephants in the Rooms Where Publics Encounter 'Science'?: A Response to Darrin Durant, 'Accounting for Expertise: Wynne and the Autonomy of the Lay Public'. *Public Understanding of Science* 17 (1): 21–33.

CHAPTER 5

Scientific Expertise and Engagement Experts

In Chap. 4, we introduced idealist and cynical purposes of public engagement. The former is highly influenced by the deliberative theories of John Rawls and Jürgen Habermas; however, using Chantal Mouffe's critiques of these two theorists, we articulated how public engagement with science events often constrains participants' capacities as a public, limits their ability to engage, and stymies "irrational" or "unreasonable" forms of expression for the sake of a democratic ideal. Using the work of Michael Warner and Robert Asen, we described how focusing on the full range of discursive and performative aspects of public engagement events opens possibilities for more ecologically valid research. The kind of data usually marginalized by democratically ideal public engagements are vernacular in character, and Samuel McCormick, Kent Ono, John Sloop, Gerard Hauser, and Robert Glenn Howard have developed distinct but related frameworks useful for researchers seeking more ecologically valid public engagement settings.

In this chapter, we offer a perspective that combats both overly cynical and idealist forms of public engagement. Lippmann's conceptualization of a "phantom" public, a public incapable of handling the complexity of modern life, is present in many public engagement with science events. Such events often assume that a public is incapable of handling complex scientific information without prior orientation by experts. These experts are usually selected by engagement experts organizing the event, and are presented to participants as specialists. Additional materials, including workbooks, handouts, or informational packets, are also designed, framed, and produced by the organizers of these events, and frequently given to

© The Author(s) 2018
A.S. Lerner, P.J. Gehrke, *Organic Public Engagement*,
https://doi.org/10.1007/978-3-319-64397-7_5

participants in the early stages of engagement to prime them for "proper" deliberation, an idealist motive that also stifles organic public engagement. Alternatively, some public engagement with science events show cynicism by participating in "downstream" engagement, or engagement that occurs well after experts have already decided a course of policy. This form of ineffectual engagement occurs when public engagement events produce no tangible policy or decision-making impacts. As Steven Emery et al. (2015) have explained, to date no systematic measure of public engagement effectiveness has been made (p. 422). Despite the increasing utilization of public engagement methods, the fact that the impact of these methods remains unmeasured demonstrates a cynical mindset. Perhaps public engagements have no measurable outcomes because researchers, organizers, and sponsors lack faith in the contributions a lay audience can make to science policy.

Engaging in a conversation with the long-standing philosophical "demarcation problem," this chapter suggests that a rhetorical standpoint can help reframe the difference between expert and nonexpert discourse and alleviate this cynical mindset. As philosopher and historian of science, Larry Laudan (1983) argues, "much of our intellectual life, and increasingly large portions of our social and political life, rest on the assumption that we…can tell the difference between science and its counterfeit" (p. 111). Continuing, Laudan (1983) explains that "for a very long time, philosophers have been regarded as the gatekeepers to the scientific estate," no doubt for their acuity at differentiating reality and mere appearance, or "real science" and "pseudo-science" (p. 111). Tracing the historical roots of the demarcation between science and nonscience reveals this divide as ephemeral at best. In ancient Greek philosophy, for instance, divisions between magic and medicine were tenuous until the rise of Parmenidean thought, which strictly demarcated "true" knowledge from "mere" opinion. Parmenides continued to influence major philosophical thought until the nineteenth century. At that point, methodology replaced epistemology as the primary condition of "true" knowledge, thus making science an extension of a certain regime of empirical procedures rather than a rare kind of knowledge. Positivists adhered to this philosophy for some time, until the popularization of Karl Popper's notion of "falsification." This influential perspective would eventually be supplanted by Thomas Kuhn and other scholars working in the tradition of the sociology of scientific knowledge. This conceptual development invites us to reconsider the authority of publics to deliberate on technical and scientific issues.

There are some important examples we can draw on to a reevaluate this public capacity. For instance, Alan Irwin (1995) has argued that "the prognosis for science, democracy and citizenship seems extremely gloomy," and in turn has proposed the possibility that we can move beyond the "current impasse in science-citizen relations" (pp. 32–33). For Irwin, this means eschewing the belief that publics are passive, homogenous, and incapable. Instead, publics should be understood as having a significant diversity of positions and "a rich pattern of knowledges and understanding" (Irwin 1995, p. 33). For example, Steven Epstein (1995) is well known for his research on how AIDS activists fought to achieve representation in the scientific process of medical research, thus demonstrating the "variety of routes by which credibility is made manifest" (p. 16). As Epstein (1995) argues, the case of AIDS research illustrates how "we find a multiplication of the successful pathways to the establishment of credibility, a diversification of the personnel beyond the formally credentialed, and hence more convoluted routes to the resolution of controversy and the construction of belief" (p. 17). This concept of the "citizen scientist" is a useful one, because it successfully buffers against the cynicism that is often found in public engagement with science events. However, much of the research on citizen science has more to do with how publics can participate in scientific projects, and less to do with how science and public policy meet each other in deliberative engagement settings. Hence, we need to look elsewhere for models that can resist the cynical mindset and avoid forcing participants into preconstructed idealist deliberative frameworks.

One such setting is the courtroom. Scientific experts and citizens have long engaged and deliberated in the context of legal cases. Indeed, science and technology studies (STS) scholars have investigated (and even participated in) these legal endeavors. Following our rough sketch of the history of demarcation, ending with the collision between Kuhn and Popper, we pick up STS scholarship on how demarcation becomes enacted in legal settings. Although there are noteworthy problems with how expertise is wielded in juridical settings, the courtroom resists the cynicism that is so often apparent in public engagement with science events. Legal decisions are deliberative, but not narrowly concerned with how the public views scientific expertise. Scientific expertise is weighed against other arguments and concerns in the courtroom, thus demonstrating a more robust representation of deliberation than many public engagement with science events. Furthermore, the demarcations between expertise and non-expertise are not unimpeachable in court as they often are in public engagement with science settings. Instead, expertise can be challenged, making demarcation not prescribed but negotiated as part of the deliberative process itself. This is a valuable model for combating the cynical mindset and avoiding precon-structed idealist "correctives" and offers a constructive version of how expertise and non-expertise can cohabitate in an ecology.

Our final task in this chapter is to briefly review and synthesize some of the arguments we have made in previous chapters and suggest that a rhe-torical sensibility can assist in producing a more ecologically valid public engagement with science event. While we are not necessarily advocating that public engagement with science experts take on the mantle of rhetorical studies, the traditional emphasis of rhetoric on context, situation, argument, and negotiation can calibrate our understanding and help reevaluate the capacity for publics to deliberate on technical topics. Following this calibra-tion, the previous chapters coalesce into the foundation for organic public engagement methods, detailed in Chaps. 6 and 7.

Proto-demarcation Perspectives

The demarcation problem is classically articulated by philosophers of sci-ence as a means to separate "deficient" modes of argumentation and rea-soning from those that seek truth and certainty. Since at least Parmenides, many ancient Greek thinkers held it important to differentiate opinion (*doxa*) and knowledge (*episteme*). Cynical purposes for public engagement assume the division between opinion and knowledge, between lay persons and scientific experts, and they do so in both their methodology and practi-cal application. Having experts guide the circulation and dispersal of dis-course, developing expert materials to facilitate certain forms of engagement over others, or establishing downstream engagements all reflect cynical purposes and demonstrate weak ecological validity. These purposes com-municate a commitment to the rigid demarcation between expert and lay knowledge. Instead of folding the process of demarcation into the delibera-tive process, sponsors of these events invent the conditions under which information is recognizable as expert as opposed to amateur. Participants are not asked to bring in their own research and arguments that they con-sider compelling, nor to help select the body of experts that guide engage-ment. In fact, many engagement events position experts as outside of the deliberative engagement entirely, and consider deliberation as occurring solely between participants. Given that individuals often participate in mul-tiple publics at once (experts are both scientists and citizens, after all), this seems to be a deeply counterintuitive practice. At their core, engagement events that participate in this demarcation appear profoundly cynical of the capacity for citizens to be agents in scientific debates, despite citizens' stakes in these matters. What these events fail to account for is the artificiality and complexity of the divide between expert and lay discourse, which is evident even in ancient pseudo-articulations of the demarcation problem.

One example can be drawn from the Hippocratic text *On the Sacred Disease*, which examines the symptoms, causes, and treatments of epilepsy. The beginning of this text is a scathing critique of "conjurors, purificators, mountebanks, and charlatans" who "referred this malady to the gods" (Hippocrates). Hippocrates condemns their superstitious cures, such as avoiding seafood "which are unwholesome to men in diseases," goat, stag, sow, or dog meat "for these are the kinds of flesh which are aptest to disorder to bowels," or even wearing black, "because black is expressive of death." However, by today's standards, Hippocrates' explanation of epilepsy is little more convincing than those he condemns: those who "have had no discharge of saliva or mucus, nor have undergone the proper purgation in the womb, these persons run the risk of being seized with this disease." As G.E.R. Lloyd (1979) suggests, the comfort Hippocrates' patients felt from his treatments were probably more psychologically motivated by their trust and confidence in his authority, rather than representing an actual means of curing epilepsy (p. 49). As Lloyd (1979) goes on to explain, no straightforward account between "science" and "philosophy" on the one hand, and "magic" and "irrationality" on the other, can be sustained given the complex relationship between and within the theory and practice of medicine on the one hand, and natural philosophy in ancient Western intellectual traditions on the other (p. 49).

Once Parmenides became a central figure in ancient Greece, this situation radically altered. Parmenides and his followers first established the foundational opposition between the senses and abstract reason, and determined their relative worth (Lloyd 1979, p. 71). In a Parmenidean model, empirical evidence and observations are relegated to mere *doxa* (opinion). Truth emerges by deduction from a single point. In turn, induction by collection of individual observations is not simply inadequate, but excluded from the rational endeavor. Plato's differentiation between the visible and the intelligible wholly plays into the Parmenidean model, whereby truth could only be known using geometry and deduction. It is not until Aristotle that we see a substantial rift in the ancient Greek philosopher's take on this distinction. For Aristotle, *episteme* denotes both a body of knowledge, and the ordering and display of that knowledge. The Aristotelian model, which essentially establishes the methodological basis of contemporary science, requires that the inquiring researcher observe nature and then seek rules to explain and predict its functioning. As Laudan (1983) explains, for Aristotle science attains its characteristic demarcation from opinion on the basis of two criteria: apodictic certainty and the differentiation between practical and theoretical knowledge (p. 112). In the former case, "the first principles of nature are directly intuited from sense; everything else worthy of the name of science follows demonstrably from these first principles" (Laudan 1983, p. 112). In the latter case, there is a strict division between practical knowledge (*techne*) and scientific or theoretical knowledge (*episteme*). Using Wynne's example of workers at a nuclear materials plant, employees knew how to handle these dangerous materials and were aware of its toxicity (*techne*), but could not differentiate different kinds of radiation, what natural forces produce radiation, or other fundamental questions that scientists posed (*episteme*).

The important lesson from this brief foray through ancient Greek scientific thought is that what counts as expert and nonexpert knowledge and discourse has no stable historical origin or referent. Lloyd (2009) illustrates how the narrow construal of what we mean by "scientific expertise" has only been practiced in the past 150 years or so, and is neither universal nor uniform in any given society at any point in history (p. 155). Hippocrates' speculations on epilepsy, grounded in the ancient system of humors, are just as incredulous to us as the "charlatans" he roundly mocks and criticizes in his work. What counts as the properly "scientific" is demarcated by Parmenides, only to be supplanted by Aristotelian induction. Here too, Aristotle's ideas did not survive untouched, as the seventeenth and eighteenth centuries saw a shift away from Aristotle's dual demarcation criteria. Most thinkers accepted Aristotle's first criterion, between empirical certainty and mere opinion, but rejected his second, between *techne* and *episteme*. Galileo Galilei, Christiaan Huygens, Isaac Newton, and others roundly disregarded the distinction between *techne* and *episteme*, and each were "prepared to regard as entirely scientific, systems of belief which laid no claim to an understanding grounded in primary causes or essences" (Laudan 1983, p. 114). Despite the divergence from Aristotle's second demarcation criteria, there was widespread agreement that scientific knowledge and apodictic knowledge were one and the same. Yet of course, this perspective would shift by the nineteenth century, demonstrating once again that the demarcation between scientific and nonscientific statements has never been an easy task.

POPPER AND KUHN'S COMPETING PERSPECTIVES

By the nineteenth century, Aristotle's demarcation criteria had been put into question. Scientific knowledge was widely accepted as imperfect, capable of amendment, and attempts at differentiating science along lines of *doxa* and *episteme* fell out of favor. Instead, scientists began to differentiate scientific knowledge on methodological grounds. Scientific knowledge was born from a rigorous set of scientific procedures that, while fallible, were still the best bet for describing and predicting phenomena. A sensible approach, some problems did emerge when it came to defining just what constituted a "scientific method." By the early twentieth century, a mix of methodological demarcation with positivism was used to suggest a foundation grounded in processes of verification, from which the boundary between science and pseudo-science could be differentiated. One such approach was philosopher of science Karl Popper's concept of falsifiability.

Deeply suspicious of induction, because "no matter how many instances of white swans we may have observed, this does not justify the conclusion that *all* swans are white," Popper (1992) developed a deductive method of testing claims called "falsifiability" (p. 4). Early in his articulation of falsifiability, he notes how shifting away from inductive reasoning puts his theory in a precarious position. Until then, induction was the best tool for differentiating science from pseudo-science. Turning the problem on its head, Popper boldly claimed that his movement *away* from inductive reason was because it lacked the capacity to suitably differentiate between scientific and nonscientific statements. Instead, for Popper (1992) the only way to properly verify the validity of a scientific statement is by the criteria of falsifiability, which exposes "to falsification, in every conceivable way, the system to be tested" (p. 20). In other words, scientific theories should be put into an arena of verification, where they are rigorously tested until proven false. Any theory that cannot be proven false is by extension, not scientific.

Taking falsifiability to its logical conclusion, or even testing its basic tenets against scientific knowledge, sometimes produces frustrating results. For instance, there are numerous scientific statements (like universal laws) that are not easily exhaustively verifiable. In particular, existential statements (i.e., that atoms exist) remain ambiguous in his system because substantial proof of their existence is difficult to produce. Alternatively, many pseudo-scientific systems of belief have aspects that are quite verifiable. For example, for over 120 years the phlogiston theory of combustion was empirically tested, theorized, and generally accepted until Antoine-Laurent Lavoisier supplanted it with an oxygen theory of combustion. In Popper's system, the belief that combustion occurs because of dephlogistication would count as a scientific argument until it came under close scrutiny and a plausible alternative was put forth. Furthermore, as David Mercer (2016) has argued, a diluted version of Popperian philosophy is sometimes utilized by opponents of the anthropogenic model of climate change, thus manufacturing doubt on the basis of falsifiability's rigorousness despite consensus to the contrary. In spite of these systematic problems, the shift toward falsifiability by Popper (and verificationism by his positivist counterparts) marks a major conceptual change for the demarcation problem. What was once a question of epistemology, or determining what counts as scientific and nonscientific knowledge, became a question of semantic strategy. In the former, to decide that a statement was scientific meant making an assessment of that statement's empirical basis in hindsight (Laudan 1983, p. 122). In the latter, scientific status "is not a matter of evidential support or belief-worthiness, for all sorts of ill-founded claims are testable and thus scientific on the new view" (Laudan 1983, p. 122).

For some, failure to demarcate scientific and nonscientific types of knowledge was not a failure at all, but a feature of scientific discovery. Kuhn's controversial and influential work *The Structure of Scientific Revolutions* is one such example. Usually remembered for his introduction of the term "paradigm shift," a concept well-trafficked outside of its original scientific context, Kuhn developed a novel perspective on science out of the work of Popper and the positivists. Until Kuhn, historians of science generally agreed that science worked on a development-by-accumulation model. In this model, scientific knowledge was built up, piece by piece, until new models and theories were developed, and so on. However, by Kuhn's (2012) time, historians began to notice how the more they studied outmoded theories, the more they realized that these theories were "neither less scientific nor more the product of human idiosyncrasy than those current today" (p. 2). This left historians with two options: if these outmoded theories are to be called "myths," then myths are producible with the same methods and persuasive for the same reasons that lead to scientific knowledge today. On the other hand, if they are to be classified as scientific, then science must include beliefs that are incompatible with the ones that we now hold. Given these options, the "historian must choose the latter" (Kuhn 2012, p. 3). As Kuhn (2012) explains, outmoded "theories are not in principle unscientific because they have been discarded" (p. 3). Thus, we needed a new historical picture of how science actually works.

For Kuhn, there is a demarcation within science between normal and revolutionary science. The former signifies "research firmly based upon one or more past scientific achievements, achievements that some particular scientific community acknowledges for a time as supplying the foundation for further practice" (Kuhn 2012, p. 10). Today, the work of normal science, or the work that the overwhelming majority of scientists engage in, is the sort of information found in textbooks, which expound a body of currently accepted theory and illustrate the applications, observations, and experiments that support the tenets of that theoretical body of knowledge. This sort of knowledge defines the boundary of any given scientific paradigm, and results gained in a particular paradigm by normal research are important because they add to the precision and scope with which a paradigmatic theory can be applied. Eventually, normal science encounters intractable problems. For instance, the phlogiston theory of combustion functioned on the assumption that when combustible materials were ignited, the phlogiston would leave that substance and thus the charred remains would weigh less. Some roasted metals actually *gain* weight when roasted, however. The increasing utilization of both the balance and pneumatic chemistry made it possible to retain the gases from reactions, which surprisingly did not detract from supporters of the phlogiston theory. Instead, they invented new explanations from within the paradigm in order to explain the phenomena they witnessed. Maybe phlogiston had a negative weight. Perhaps outside particles swapped places with phlogiston when a material is roasted. More effort and focus becomes devoted to these sorts of problems in a given paradigm when faced with a crisis, and normal science continues onwards, dedicating its efforts to expanding the scope and precision with which a particular theory is applied.

Revolutionary science begins with this sort of crisis and plays off the increasing confusion that faces normal scientific activity. As Kuhn argues, there are only two universal effects of a scientific crisis. First, "all crises begin with the blurring of a paradigm and the consequent loosening of the rules for normal research" (Kuhn 2012, p. 84). Second, all crises close in one of three variations: normal science proves capable of handling the problems, the problem persists regardless of new approaches, or the crisis ends with the emergence of a new paradigmatic candidate. This third result is a "scientific revolution," and is characterized as "non-cumulative developmental episodes in which an older paradigm is replaced in whole or in part by an incompatible new one" (Kuhn 2012, p. 92). Scientific revolutions result in more of a shift than a total upheaval in understanding, primarily by explaining both well-situated and tested theories and problems that the current paradigm has trouble clarifying.

Kuhn exemplifies a major shift in conceiving the demarcation problem and suggests that the boundaries between science and nonscience are not problematic at all, but a characteristic feature necessary for understanding how science has developed throughout history. Unlike Popper, who introduced an element of contestability to further solidify the demarcation between science and nonscience, Kuhn suggested that clashing incommensurate ideas is the condition under which new and revolutionary scientific changes occur. Intractable problems force scientists to reevaluate long-standing assumptions. Crises provide the basis for further scientific development. Public engagement events rarely account for or simulate these features of science into the design of public engagement events. While it is unlikely that public engagement participants will upend long-standing scientific paradigms, engagement events frame science as an already concluded endeavor by having preestablished and unchallengeable boundaries of expertise. As Simon Locke (1999) puts it, this view "implies an idealized vision of scientific enquiry as internally harmonious, self-consistent, and resolved around its empirical, methodological and theoretical substance" (p. 76). By extension, nonscientific enquiry is delegitimized as being chaotic, inconsistent, and lacking any stable substance. This cynical standpoint perpetuates a version of science in engagement events that lacks ecological validity because it does not capture the full scope of activities in which scientists actually engage.

Popper and Kuhn's contributions stand out as central pivot points for the demarcation discussion. They both demonstrate how science is a contestable ground and push to deny or embrace that demarcation between science and nonscience has long been an important facet of scientific practice. We wish to be clear on one point, however. While it may seem that both the failure of Popper and the radical reevaluation of Kuhn lend themselves to putting scientific knowledge in question, we prefer to think of these shifts as less about the potential failures of science and more about the potential successes of nonscience. Science is without a doubt the best tool available to make predictions about our world, and the work that scientists do is both valuable and demonstrably successful. Moving past a perception of a "deficit" between publics and scientists (a cynical position) was a key impetus for the development of more engaging and deliberative models of science communication. But these noble efforts are doomed to fail if they do not take seriously a given publics' position, regardless of how poorly that position may fit into the current paradigm. By ignoring these positions and artificially framing expertise, the data that emerges from public engagement with science events is more a reflection of how well an event can inculcate a public into a scientific paradigm than it is a reflection of an actual public's position on a scientific topic. It follows that these events lack ecological validity on two registers: they distort a given public's position and they present a distorted version of scientific certainty. If public engagement with science experts wish to remedy these conditions and generate ecologically valid data, they must account for these distortions.

DEMARCATION GOES TO TRIAL

In the court of law, making decisions that do not necessarily align with scientific proof is part of the deliberative process, and thus can be instructional in developing a more robust version of public engagement where demarcation is more procedural than artificially *ad hoc*. Scientific information is always already caught up in an ecology, and in court cases this ecology is weighed against other factors in the deliberative process. In deliberative engagement events, however, lived ecologies are constrained for both cynical and idealist purposes and strictly demarcated at the outset of engagement. Studies at the intersection of STS and the law are a particularly valuable site for understanding how demarcation is situated in a particular ecology, and thus instructive for making a case for more ecologically valid engagement practices. As Sheila Jasanoff (1995) has argued, the "law has devised a complex system of rules and practices for choosing what to believe when facts are uncertain" (p. 10). This complex system of rules helps define the boundaries of what counts and does not count as "expert" testimony, and this procedural mechanism functions in no way "scientifically." As Jasanoff (1995) explains, the "legal system's allegiance to values other than those of science may open the way to decisions that look like sheer irrationality;" however, sometimes this "irrational" refusal to accept well-supported scientific evidence (like in the case of Charlie Chaplin's paternity case) could be "characterized as social wisdom rather than scientific illiteracy" (p. 11). Unlike public engagement with science events, where expert scientific information is rarefied and artificially demarcated from other types of data, legal settings demonstrate how scientific information can function in a deliberative setting without overruling the possibility of dissent.

In a court of law, expertise is a negotiated feature of the deliberative process. While scientific evidence and expertise is no doubt a valuable persuasive tool in legal settings, the blend of ethical, moral, and practical concerns in a legal case are weighed together with the credibility of scientific information. For instance, Gary Edmond and David Mercer (1997) brilliantly explored US litigation involving birth defects caused by the drug Bendectin. They provided a sophisticated analysis of a jury's capacity to appropriately utilize technical and scientific information in rendering a verdict. Edmond and Mercer (1997) argue that Popperian and positivist approaches to science, or approaches that are interested in the "correct" understanding of scientific information, have been "nurtured by concerns among scientific organizations and industry lobby groups that there has been a decline in their social authority…because of the failure in the public understanding of science" (339). They argue these approaches are simplistic and common in legal settings. Even with these rigid and reductionist demarcations in place, Edmond and Mercer (1997) adopt a perspective common to sociology of scientific knowledge, and conclude that in practice "there is no simple epistemological basis on which competence may be determined" in the proceedings of a legal case (p. 349). In a later study, Edmond (1998) explored responses to scientific evidence from a lay person accused of murder, where he continued to argue that the expert and lay knowledges "often coalesce in ways that are rarely either mutually exclusive or divorced from the social worlds and previous experiences of the protagonists" (p. 106).

A second study of scientific evidence used in juries took a very different approach to exploring demarcation in the legal setting. In this study, Michael Lynch and Simon Cole (2005) examined the problems that Cole encountered when appearing as an expert witness in the October 2001 case *People v. Hyatt*. Here, Cole was asked to remark on the reliability of fingerprints. Forensic science's reliance on fingerprinting was an invention of the early twentieth century, and was considered an "infallible" source of expert evidence in criminal trials. This of course assumes that fingerprint examiners could infallibly find matches between fingerprints. However, in the 1990s there were two cases that upended the unquestioned reliability of fingerprints: *Daubert v. Merrell Dow Pharmaceuticals, Inc.* and a further clarifying ruling in *Kumho Tire Co. v. Carmichael*, both of which altered admissibility protocols for expert evidence.[1] Importantly, these rulings introduced the requirement for pretrial admissibility hearings to assess the "general acceptance" of a theory in a given scientific field. As Lynch and Cole (2005) note, this "posed obvious problems, as it left open the question of how to constitute the 'relevant scientific community' as well as how to distinguish closed communities of true believers from established scientific fields" (p. 271). With the rise of DNA testing, this made it possible to argue that "(1) fingerprinting had, for historical reasons, escaped commensurate scrutiny; and (2) fingerprinting had claimed greater evidentiary value (absolute certainty) based on a thinner scientific foundation" (Lynch and Cole 2005, p. 272). While Cole was originally asked to remark on the admissibility of fingerprint evidence, in an unusual turn of events Cole was invited to be held accountable for his own expertise about science. This provided an opportunity for Cole to undergo questioning concerning his ability to demarcate scientific knowledge and practice. This cross-examination produced a precarious reflexive position that forced Cole to contend with the precarity of his expertise as a STS scholar. Although the court found the fingerprints admissible as evidence, Cole's constructivist, non-positivist stance toward science was a novel perspective in the strictly demarcated legal setting. The deliberative mechanism of cross-examination put him in a precarious position, but this position rightly helped demarcate his own status as expert. Thus, legal proceedings function by not only weighing scientific factors against the larger ecology of relevant concerns but also have the capacity to rigorously monitor and challenge those who possess the authority to decide what is and is not scientific evidence.

One of the more contentious court cases involving STS came only a few years after Lynch and Cole's examination of *People v. Hyatt*. In this case, *Kitzmiller v. Dover Area School District*, Steven Fuller was asked to defend the inclusion of "intelligent design theory" alongside the neo-Darwinian theory of evolution in the Dover area public school system. Fuller's rhetorical strategy was to undermine the view that neo-Darwinism was an uncontested field of scientific research. His argument was not an admonition of intelligent design theory, but a principled stance against the incontestability of Darwinism. As Fuller (2006a) explained, his "own view is that the defense did indeed have the weaker case, but equally, that the judge did an injustice to the relevant philosophy, politics, and ultimately to science" (p. 830). In contrast to Cole, who was put into a rhetorically reflexive position with regard to his own authority, Fuller's contradictory argumentative position is never thoroughly examined during the case (Edmond and Mercer 2006, p. 844). Without the deliberative mechanism of cross-examination,[2] the view that science is socially constructed can be wielded to undermine any scientific claim without impedance. The constructivist argument can function to short-circuit the *ethos* of any given scientific claim to expertise, since it destabilizes the difference between facts and opinions. As Lynch (2006) explains, Fuller could have made an argument for either side of the *Kitzmiller* case. The same thing could be said of the *People v. Hyatt* case, where Cole could have suggested that while fingerprint analysis is not a science *per se*, a "proponent of STS could easily endorse fingerprinting as a practical art that relies upon the unacknowledged skills of technicians, and the crucial role of tacit knowledge" (Lynch 2006, p. 824). In fact, Lynch (2006) notes how it is "more consistent with STS orthodoxy to take the fingerprint examiners' side rather than to support the side that Cole takes" (p. 824). This multidimensional rhetorical mutability does not make the social constructivist approach all-powerful; because the legal setting positions expertise in a larger deliberative ecology, different forms of argumentation are able to jockey for convincingness. As in Cole's case, the capacity to blur opinion and fact is self-defeating when tested.

These jury studies demonstrate two major practical takeaways for public engagement with science experts. On the one hand, they provide examples of how expertise is negotiated in a deliberative system. On the other hand, they offer a buffer against the critique that blurring the line between experts and engagement participants will take a science engagement event away from its intended purpose of fostering a relationship between the work of professional scientists and publics. In the first case, as Edmond and Mercer note in their study, legal standards for delineating scientific from nonscientific information often demarcate experts and nonexperts along rigid boundaries. While expertise surely has more rhetorical force in such a demarcated system, this force is not unlimited in its capacity for persuasion. This is because demarcation is itself a negotiated and contested aspect of the legal ecology in which it is embedded. Engagement events do not include this mechanism. Instead, sponsors generally dictate demarcation before deliberation begins. This approach to public engagement assumes one of two cynical scenarios: either the public is incapable of negotiating demarcation as part of the process of deliberation, or experts are incapable of supporting their arguments when faced with potential criticisms. By including a negotiated demarcation as part of the deliberative process, no single persuasive appeal will dominate the engagement ecology, thus eliminating both problematic situations. Reevaluating the position of a public's expertise does not mean automatically valuing any and all arguments, opinions, and positions held by those publics. It means allowing a public's persuasive capacity and vernacular expertise to coexist in the larger ecology of deliberation without artificial constraints. When positioned in this larger ecology (like in a court of law), demarcation becomes a dynamic organic process instead of a static prescription. As we have argued in Chap. 4, ecological validity cannot be derived from an engagement event that constrains deliberation for either idealist or cynical purposes. The latter signifies that participants are not capable engaging with how the process of demarcation occurs in an ecology. Instead, publics are guided toward certain ways of arguing, certain forms of evidence, or certain frameworks of expertise that from the outset are uncontestable. Recognizing that science is in part constructed through social processes does not mean that all positions are equally valid, that all forms of evidence are equally persuasive, or that all claims are ethically justified; it all depends on the ecology.

FINDING RHETORIC IN THE THIRD WAVE

Cynical engagement purposes do have an intelligible position, and in order to dispense completely with this position, we need to further contest potential critiques against a reevaluation of vernacular expertise. There is a very real danger in lowering the bar for what counts as "proper" modes of deliberation. In Fuller's case, he ended up defending a theory that is not rigorously tested nor widely believed by experts and most publics. Alternatively, according to the Centers for Disease Control and Prevention, 2014 saw the largest number of US measles cases since measles elimination was documented in the United States in 2000. This outbreak was largely due to the popularity of the anti-vaccinationist movement, which is widely discredited by professional medical and scientific communities. Imagine a public engagement event centered around nanomedicine. It is not unreasonable to think that some of the participants in this event might hold pseudo-scientific beliefs about medicine. What if one of these participants is particularly persuasive, and derails the discussion by suggesting that nanomedicine will interfere with an alternative therapy, such as crystal healing? Should sponsors intervene and "correct" this participant? Should the crystal healer's claims be ignored? Since crystal healing and other non-institutional medical beliefs are actually held by publics, it follows that intervening will interfere with the ecology, thus lowering the ecological validity of the engagement event. Furthermore, totally ignoring these sorts of claims risks distorting our analysis of the data, since we are artificially excluding an aspect of how a public actually encounters scientific information. On the other hand, letting this suspect argumentative position sway other participants may risk undermining the implicit goals of public engagement events or detract from deliberation on the topic at hand.

This basic issue is at the heart of Harry Collins and Robert Evans's influential essay, "The Third Wave of Science Studies." As they explain, one of sociology of scientific knowledge's major contributions has been to demonstrate that scientific information has no special access to truth (Collins and Evans 2002). For Collins and Evans, it logically follows that since it is no longer clear that scientists have special access to the truth, then it is equally unclear that we should specially value their advice. This issue has a strong resonance with our claim about public engagement with science events. As Collins and Evans (2002) explain: "Should the political legitimacy of technical decisions in the public domain be maximized by referring them to the widest democratic process, or should such decisions be based on the best expert advice?" (p. 235). Public engagement with science events act as if they are performing the former, but by virtue of their procedural construction, usually only accomplish the latter. In the first case, we risk including all sorts of "crackpot" science. In the second case, we risk stripping deliberation of any real democratic character, thus defeating the entire purpose of public engagement events in the first place. As Collins and Evans argue, this tension is between what they call the "Problem of Legitimacy" and the "Problem of Extension." They argue STS has largely resolved the problem of legitimacy by "showing that the basis of technical decision-making can and should be widened beyond the core of certified experts" (Collins and Evans 2002, p. 237). They also suggest that STS has failed at accounting for the problem of extension, which asks how far participation in technical deliberation should go. In order to sort out how STS got to this position and where it should go in the future, they outline three "waves" of STS.

The first wave is pre-Kuhn, and fits into the conversation that Popper and the positivists inherited from the long history of the demarcation problem in Western thought. In this period, the problem of extension does not exist, because scholars were largely unaware that a problem of legitimacy existed. The second wave, which runs post-Kuhn until today, is the period we have outlined above which consists of social constructivism, or more specifically, the sociology of scientific knowledge. For Collins and Evans (2002), wave two denotes a period where science is reimagined as a social phenomenon (p. 239). Hence, the boundaries of what can count as expertise become enlarged, thus opening the door for the problem of extension. This problem is readily viewable in the court cases that we have discussed in this chapter thus far: in the case of Cole, his reticence about his own authority is a natural outgrowth of an incapacity to determine how authority is ascribed to experts. In Fuller's case, his unchecked authority allows the constructivist position to be extended to untenable arguments and used for political purposes that extend beyond the outlines of science. For Collins and Evans (2002), the third wave attempts to move the discussion away from epistemology and toward studies of expertise and experience, or "SEE" for short. Building off the research found in wave two, Collins and Evans endeavor to construct a normative theory of expertise that is compatible with the social constructivist position. The third wave of science studies is, in a way, a return to the boundaries of wave one, with the added knowledge that these boundaries are not innate to scientific practice, but instead artificially and socially constructed borders useful for solving the problem of extension.

One problem with Collins and Evans's characterization of STS and their subsequent move toward a normative basis for expertise is that it obscures the complex ecological context in which scientific information meets up with publics. As Wynne (2003) argues, Collins and Evans "seem to have confused the narrow confines of laboratory sciences [studies] with the fundamentally different further issues raised for scientific knowledge in public arenas" (p. 404). As Wynne suggests, Collins and Evans seem to characterize their normative position grounded in philosophy of propositionalism, an epistemic model where every intentional attitude is a proposition about some object, event, idea, and so on. However, citing studies of public policy and risk management decisions, Wynne (2003) notes how public contestation of science is more about a public's understanding and definition of the issue at hand than it is about propositional truths (p. 404). In other words, Wynne argues that Collins and Evans seem to misunderstand how scientific practices get articulated in public ecologies, which always contain propositional statements, (i.e., "are nanomachines dangerous?" or "should we label GMOs?") within a broader ecological context where meanings and definitions are in a state of flux. Wynne (2003) concludes that their position "corresponds with a neglect of context and a denial of the ultimate contingency of saliency and meaning" (p. 404).

Another critique of Collins and Evans's "Third Wave" is that it mischaracterizes and unwittingly distorts the goals of public engagement. As Jasanoff (2003) cleverly puts it, "they are trying to lock the barn door after the horse has already bolted" (p. 397). While Collins and Evans perceive the fundamental theoretical and philosophical problems associated with the problem of extension, their push to draw boundaries to limit the scope of public participation misses the fact that most Western states have come to grips with the idea that publics are responsible enough to decide how and to what extent they desire to engage in decision-making (Jasanoff 2003, p. 397). Jasanoff (2003) explains how new areas of technical deliberation are emerging across the world, "where the rules of participation are not yet laid down and the pulls of democracy on behalf of a global civil society are tenuous at best" (p. 397). Thus, creating restrictive categories that limit the capacity for citizens to adjudicate technical decisions may be theoretically warranted, but practically unnecessary if this capacity is always already being negotiated organically in deliberative democratic experiments. There is no need to artificially demarcate expertise from lay knowledge, because these "demarcations will keep being produced in any case, in the everyday work of scientists, citizens and institutions of governance" (Jasanoff 2003, p. 398). We would only add that this depends upon engagement events allowing this everyday work of scientists and citizens to actually occur.

Darrin Durant (2011) has characterized the debate between Collins and Evans on the one hand, and Wynne and Jasanoff on the other, as being shaped by alternative takes on liberal democratic theory (p. 692). As he explains it, Collins and Evans seem to take Rawls's concept of public reason and liberal egalitarianism to heart, while Wynne and Jasanoff align with identity politics specifically, and Habermas's discourse ethics more generally (Durant 2011, p. 692). Durant comes to the aid of Collins and Evans in order to buffer them from claims (like Jasanoff's) that the Third Wave is "anti-democratic" because it proposes limits on a public's capacity to deliberate. As he explains, if "politics is about prescribing the kind of argument capable of facilitating legitimate consensus, then Jasanoff and Wynne's participatory politics leaves relatively unanswered what to do with intractable social conflict" (Durant 2011, p. 692). While Durant's characterization of these positions might very well be accurate, as we have argued in Chap. 4, in practice these political theories never quite become reified in public engagement events, and even if they did, they pose different sets of practical problems that are more broadly associated with idealist modes of constructing actual deliberative events. For public engagement with science, the problem is not as Durant characterizes it here, between Collins and Evans's Rawlsian approach on the one hand and Wynne and Jasanoff's Habermasian influence on the other. Rather, the problem is that the theoretical posturing of all these positions seems to get lost in translation to organized, actually occurring deliberative events.

We take a different approach. In order to produce more ecologically valid research and engagement events, what matters is not the epistemological or ontological status of expert and nonexpert knowledge. What matters is the capacity to *rhetorically negotiate the status of expertise and non-expertise.* We have produced ample evidence that demarcation is a historically mutable rhetorical exercise, where depending on who is making the claims, what evidence they are marshaling, what techniques they are using, and how they are presenting their arguments, demarcation takes on certain forms. In other words, demarcation is always caught up in everyday rhetorical ecologies. The restrictions imposed by both idealist and cynical purposes for public engagement lead to the same consequence: the capacity for negotiation, debate, persuasion (put simply, *rhetoric*) is artificially constrained, thus making engagement events lack ecological validity. They lack ecological validity because democracy, politics, and everyday life are permeated by subtle and not-so-subtle rhetorical sensibilities. Allowing vernacular data to appear and recording it for analysis is a step toward stronger ecological validity, because as it stands now, the only group possessing the means to persuade are sponsors, while participants are only expected to be persuaded. This is not engagement, it is propaganda.

Putting vernacular data on the same level as expert data does not mean that these two necessarily share fundamental characteristics. Rather, putting them on the same level means allowing them to engage one another rhetorically. While experts might indeed have the upper hand in public engagement events, ruling out the possibility for other rhetorically effective arguments or constraining the capacity for participants to utilize their own rhetorical force is not simply undemocratic, it does not reflect the reality of how publics actually engage with science. Publics concerned about the potential for vaccinations to give their children autism are not simply outside the boundaries of rationality and therefore on exclusionary deliberative grounds. These publics are acting in order to protect their children, or have reason to distrust the scientific institutions, or are persuaded by other logical, emotional, or authoritative arguments that are more or less valid. Without letting these arguments coexist in the ecology of public engagement events where scientific claims and expertise go largely unchallenged, we are not staging engagement events that reflect actual practices of deliberation "in the wild." We do so at our own peril. While a rhetorical turn for public engagement with science surely presents its own problems, at the very least it shakes up this debate "mired in an unproductive stand-off about the scope of politics" (Durant 2011, p. 692).

RHETORICAL ECOLOGIES AND THE TURN TOWARD ORGANIC PUBLIC ENGAGEMENT

S. Scott Graham (2015) has recently argued that while the "history of STS inquiry offers a great number of tools that can assist in the investigation" of scientific topics, "STS is still not enough" (p. 8). Both Wynne and Jasanoff have suggested that context is a crucial aspect of public engagement with science, and rhetoric has long contributed a great deal to better understanding how representation and meaning circulate in context. As Graham (2015) explains, "the notion of context – alongside audience and purpose – is typically identified by rhetoricians as one of the core features of any moment of discourse" (p. 8). While definitions of "context" will differ between rhetorical studies scholars, the core of this belief comes from one of the first formal treatments of rhetoric by Aristotle. As he defines it in book 1, chapter 2 of the *Rhetoric*, rhetoric is the capacity to see, *in each given case*, the available means of persuasion. Some rhetorical principles may have a vague universal applicability, for instance, that emotion, logic, and credibility are persuasive, but overall, rhetoricians know full well that rhetoric begins with the case at hand and builds out a theory from the operant context. Rhetorical study's emphasis on context speaks to the concerns of scholars like Wynne, who have long argued that demarcation is built out of the context in which it appears.

Although rhetorical studies have yet to be applied to public engagement with science events, its presence in STS is neither foreign nor novel. For instance, Fuller (2006b) has called for engagements with rhetorical scholars in communication studies across American campuses (p. 174). Alternatively, Alan Gross (2006) is a recognizable name in many STS circles for his important contributions toward the rhetoric of science.

Furthermore, rhetorical scholars Paul Lynch and Nathaniel Rivers (2015) have published an edited collection of essays all relating to the work of Bruno Latour, who has deeply affected certain strands of thinking in rhetoric and composition. There are numerous more examples that could be presented, but it should suffice to say that rhetoric and STS have largely been complementary fields of study for some time. As Graham suggests however, the most important contemporary change that has aligned STS with rhetorical studies has been a shift toward "new materialisms." Diana Coole and Samantha Frost's *New Materialisms* marked this shift most clearly in 2010, where they collected a series of essays from scholars across several different disciplines who were interested in incorporating a more "ontological" approach in their analyses. One artificial constraint of rhetorical studies has long been its emphasis on texts and textuality. However, the move toward new materialism marked an opening for rhetorical scholars to see beyond the text, and thus move away from traditional rhetorical inquiries using classical tropes, figures, and schemas. This, in turn, shifted rhetorical studies toward an approach more closely resembling the work already done by sociology of scientific knowledge scholars and other sociologists who work in and around STS. As Graham (2015) has argued, this shift has created a situation where "scholars of rhetoric and STS have reached an exciting moment of theoretical symmetry" (p. 13).

The "materialist" turn in rhetoric can also be described as an ecological turn, which fits neatly into the argument we have made in Chap. 2. Jenny Edbauer's (2005) essay, "Unframing Models of Public Distribution" reevaluates the basic foundations of rhetorical analysis from rhetorical situations (which focuses on audience, exigence, and constraints) to "lived practical consciousness or structures of feeling" (p. 5). Her articulation of rhetorical ecologies in many ways mirrors the phenomenological basis that undergirds the contours of ecological thinking we have already traced. In addition, she incorporates Michael Warner's conception of publics into her description of ecology, thus highlighting how his conception of publics extends beyond a narrow discursive reading to a wider ecology of rhetorical meaning. However, the most thorough extension of ecological thinking derives primarily from the work of Thomas Rickert (2013), whose *Ambient Rhetoric* picks up strains of phenomenological arguments from Martin Heidegger as well as addresses Latour's "actor-network theory." Here, he situates rhetoric in relation to the Heideggerian concept of "dwelling," which describes the mode of existence that we have already defined through the works of James Gibson, Kurt Lewin, and other prominent ecological thinkers. Extending Graham's basic argument that STS and rhetoric have reached "theoretical symmetry," the works of Edbauer and Rickert are exemplars of how this turn toward "materialism" is truly a turn toward ecological thinking. Recall that ecological thinking is an additive principle. For rhetoric, this means including extra-textual and non-discursive elements into analysis. For public engagement with science experts, practitioners, and theorists, this means including a rhetorical sensibility that opens the floor for more organic modes of deliberation, modes of deliberation constrained by neither idealist or cynical purposes of public engagement.

We are not suggesting that public engagement with science scholars suddenly start performing rhetorical analyses of engagement events. Rather, we are suggesting they take a contemporary rhetorical mindset when approaching both the construction and analysis of public engagement events. Artificially constraining and manipulating the functionality of public engagement events results in rhetorical blockages that in turn prevent publics from functioning organically, as we have argued in Chap. 4. Alternatively, not accounting for or not allowing vernacular data to emerge in engagement events also limits the possibility of rhetorical circulation, and thus potentially weakens and obscures an entire ecology from being accurately represented in research. Sophisticated rhetorical analysis begins with the case at hand and builds out a theory, methodology, or argument from that case. In turn, if public engagement with science events wish to be more ecologically valid, they too should begin with sites where citizens interested in engaging with science are located and build out an engagement methodology organically. This does not mean plucking them from their lived ecologies. This does not mean sending them a survey with limited choice. This does not mean analyzing media that targets these publics and inferring public attitudes from these analyses. This means maintaining the autonomy and ecology in which publics circulate meaning. To do so is to perform an ecologically valid engagement event. We call this *organic public engagement*.

NOTES

1. The most authoritative and superlative overview of this case can be located in Jasanoff's *Science at the Bar*.
2. More specifically the *Daubert* challenges, which is a hodgepodge of social constructivist, Popperian, and positivist benchmarks for determining expertise.

REFERENCES

Collins, H.M., and Robert Evans. 2002. The Third Wave of Science Studies: Studies of Expertise and Experience. *Social Studies of Science* 32 (2): 235–296.

Durant, Darrin. 2011. Models of Democracy in Social Studies of Science. *Social Studies of Science* 41 (5): 691–714.

Edbauer, Jenny. 2005. Unframing Models of Public Distribution: From Rhetorical Situation to Rhetorical Ecologies. *Rhetoric Society Quarterly* 35 (4): 5–24.

Edmond, Gary. 1998. *Down by Science*: Context and Commitment in the Lay Response to Incriminating Scientific Evidence During a Murder Trial. *Public Understanding of Science* 7 (2): 83–111.

Edmond, Gary, and David Mercer. 1997. Scientific Literacy and the Jury: Reconsidering Jury 'Competence'. *Public Understanding of Science* 6 (4): 329–357.

———. 2006. Anti-social Epistemologies. *Social Studies of Science* 36 (6): 843–853.

Emery, Steven B., Henk A.J. Mulder, and Lynn J. Frewer. 2015. Maximizing the Policy Impacts of Public Engagement: A European Study. *Science, Technology, & Human Values* 40 (3): 421–444.

Epstein, Steven. 1995. Democracy, Expertise, and AIDS Treatment Activism. In *Science, Technology, & Democracy*, ed. Daniel Lee Kleinman, 15–32. New York: State University of New York Press.

Fuller, Steven. 2006a. A Step Toward the Legalization of Science Studies. *Social Studies of Science* 36 (6): 827–834.

———. 2006b. *The Philosophy of Science and Technology Studies*. New York: Routledge.

Graham, S. Scott. 2015. *The Politics of Pain Medicine: A Rhetorical-Ontological Inquiry*. Chicago: University of Chicago Press.

Gross, Alan G. 2006. *Starring the Text: The Place of Rhetoric in Science Studies*. Carbondale: Southern Illinois University Press.

Hippocrates. On the Sacred Disease. *The Internet Classics Archive*. http://classics.mit.edu/Hippocrates/sacred.html.

Irwin, Alan. 1995. *Citizen Science: A Study of People and Expertise and Sustainable Development*. New York: Routledge.

Jasanoff, Sheila. 1995. *Science at the Bar*. Cambridge: Harvard University Press.

———. 2003. Breaking Waves in Science Studies: Common on H.M. Collins and Robert Evans, 'The Third Wave of Science Studies'. *Social Studies of Science* 33 (3): 389–400.

Kuhn, Thomas. 2012. *The Structure of Scientific Revolutions*. 50th anniversary ed. Chicago: University of Chicago Press.

Laudan, Larry. 1983. The Demise of the Demarcation Problem. In *Physics, Philosophy and Psychoanalysis*, ed. R.S. Cohen and L. Laudan, 111–127. Dordrecht: Reidel.

Lloyd, G.E.R. 1979. *Magic, Reason and Experience: Studies in the Origin and Development of Greek Science.* Cambridge: Cambridge University Press.
———. 2009. *Disciplines in the Making: Cross-Cultural Perspectives on Elites, Learning, and Innovation.* Oxford: Oxford University Press.
Locke, Simon. 1999. Golem Science and the Public Understanding of Science: From Deficit to Dilemma. *Public Understanding of Science* 8 (2): 75–92.
Lynch, Michael. 2006. From Ruse to Farce. *Social Studies of Science* 36 (6): 819–826.
Lynch, Michael, and Simon Cole. 2005. Science and Technology Studies on Trial: Dilemmas of Expertise. *Social Studies of Science* 35 (2): 269–311.
Lynch, Paul, and Nathaniel Rivers, eds. 2015. *Thinking with Bruno Latour in Rhetoric and Composition.* Carbondale: Southern Illinois University Press.
Measles Cases and Outbreaks. *Centers for Disease Control and Prevention.* http://www.cdc.gov/measles/cases-outbreaks.html.
Mercer, David. 2016. Why Popper Can't Resolve the Debate Over Global Warming: Problems with the Uses of Philosophy of Science in the Media and Public Framing of the Science of Global Warming. *Public Understanding of Science*: 1–14.
Popper, Karl. 1992. *The Logic of Scientific Discovery.* New York: Routledge.
Rickert, Thomas. 2013. *Ambient Rhetoric: The Attunements of Rhetorical Being.* Pittsburgh: University of Pittsburgh Press.
Wynne, Brian. 2003. Seasick on the Third Wave? Subverting the Hegemony of Propositionalism. *Social Studies of Science* 33 (3): 401–417.

CHAPTER 6

Organic Engagement of Public Ecologies

In the first chapter, we explored the complex and varied development of an ecological mindset and methodology. Expanding on this intellectual history, practitioners using organic public engagement can achieve stronger ecological validity by minimizing disruption of the public ecologies they study. As Lewin rightly argued, ecologies are not limitless systems, but living worlds composed of human and non-human elements that are coherent insofar as they are meaningful. Publics are complex matrices of circulating discourses, performatives, and meanings; to attend to how these elements relate to each other and the environment in which they circulate is to perform more ecologically valid research. Public engagement practitioners should strive to replicate the conditions under which actual behavior occurs within public ecologies in order to achieve higher levels of ecological validity. For public engagement with science, this means recording and analyzing where publics deliberate on scientific topics instead of plucking them out of their native ecologies, inferring their opinions through media studies, or limiting the context of their attitudes through the use of surveys.

Chapter 3 explored why some common public engagement methodologies fall short of producing ecologically valid results. For instance, when studying conditions of anonymity, privacy, aggregation, isolation, and forced limited choice, surveys can potentially deliver ecologically valid data. However, surveys are also relatively poor at representing how publics collectively respond to new information, how publics form opinions, or

© The Author(s) 2018
A.S. Lerner, P.J. Gehrke, *Organic Public Engagement*,
https://doi.org/10.1007/978-3-319-64397-7_6

how publics are persuaded. The deficit model, which is largely supported by survey studies, contends that "the public" lacks literacy in scientific topics. Their results do not explain why publics lack scientific literacy, how they use scientific literacy in their everyday lives, or what scientific literacy means to their community. Therefore, these surveys have weak ecological validity. Alternatively, media studies generally build up a representation of the information reaching a public, but at best, they can only speak to journalistic and editorial ecologies that produce these messages, offering little to elucidate the behavior of existing publics. Researchers' interpretations of media portrayals or science journalism may not reveal how publics actually view, understand, and deploy these forms of media in their lived ecologies, thus media studies often lack strong ecological validity in relation to an actually existing public.

The shift from public understanding to public engagement with science, while an encouraging development, still tends to develop engagement events with weak ecological validity. Most deliberative engagement events have an experimental design that drives researchers toward artificial deliberative contexts. These events often prioritize both efficiency and control: the former because there are limited resources to perform public engagement research and the latter because it assures that the outcome of the experiment will be viable for publication or to appease funding agencies. In Chap. 4, we explored how controlling public engagement events for the sake of idealist purposes (purposes that lead to "proper" modes of deliberation) weaken ecological validity because they artificially constrain publics' activities. In Chap. 5, we examined how controlling public engagement events for the sake of cynical purposes (purposes that "save" publics from themselves) weaken ecological validity for similar reasons. In Chap. 4, our turn to Warner's conceptualization of publics, as well as Ono, Sloop, Hauser, Howard, and McCormick's emphasis on vernacular rhetorics, ecologically recalibrated our understanding of a public's reaction to scientific information. Alternatively, Chap. 5's general call for a more rhetorically oriented perspective toward public engagement with science research also recalibrated our thinking for ecologically valid research.

In this chapter, we articulate a public engagement methodology that carefully navigates the middle ground between idealist and cynical purposes, combining Warner's understanding of publics with vernacular rhetoric and ethnography, leading to a coherent set of principles for more ecologically valid research, which we call "organic public engagement."

THE "QUASI-ETHNOGRAPHIC" METHODOLOGY

To retain as much of the ecology of a public as possible, organic public engagement begins by developing an understanding of the public being engaged and adapts its methodology to suit that public's ecology. The design of organic engagement events and research bears in mind that, as Wynne (1991) found, people encounter and process information, including scientific knowledge, "as part of their concrete experience and position in particular institutional processes" (p. 115). To borrow from Dwight Conquergood (1992), the ethnographic disposition of being "open, flexible, adaptable, and sensitive to situation, circumstance, and nuance" is both essential to good ethnographic research and vital for developing more ecologically valid research (p. 81). Including ethnography into our discussion is therefore critical for understanding organic public engagement.

Scholars in vernacular rhetoric and ethnography have recognized their common interests for two decades. Sloop and Ono (1997) noted the value of ethnography for the study of vernacular rhetoric, while Conquergood (1992) argued that ethnography shares important intersections with rhetoric and communication studies. Likewise, some public engagement researchers note the potential value of ethnographic methods.[1] Organic public engagement, focused on vernacular discourses of existing publics, deploying an ethnographic sensibility, reflects Aaron Cicourel's (2007) recommendations for methodologies with stronger ecological validity: "direct observation of and participation in the everyday activities or practices of human actors in their daily life experiences" (p. 735). This might well characterize Brian Wynne's work with sheep farmers in Cumbria, John Stone's work on public participation in Great Lakes environmental management, and Annemarie Mol's work on atherosclerosis. Wynne, Stone, and Mol each use a "quasi-ethnographic" approach to their respective research sites.

Wynne's (1992) analysis of Cumbrian sheep farmers illustrates how a quasi-ethnographic disposition can produce more ecologically valid data, in part because he both spends time understanding the lived ecology of these farmers and places their discourse within the broader context of their history, skills, and locality. Following the Chernobyl disaster, some areas of Britain were contaminated by variable deposits of radioactive isotopes from localized weather patterns. At first, the effects of this contamination were dismissed by experts and politicians as negligible, but after a brief period, a ban was instituted on the slaughter and movement of sheep in the affected areas. The farmers were rightly shocked by this sweeping declaration, but reassured by experts that the elevated levels of radiation found in their flocks of sheep would last less than a month, and at the end of this period they would be allowed to sell their sheep as normal. However, the ban was later made indefinite, threatening the wholesale slaughter and ruin of these sheep farmers' livelihoods.

By chance, this area was occupied by the Sellafield nuclear complex, which does nuclear materials processing and storage functions and contains nuclear reactors, chemical reprocessing plants, and plutonium storage facilities. The biggest employer in the area at the time, this complex dominated the region socially, culturally, and economically. Coincidentally, a small wedge of land with high exposure from the Chernobyl-related weather patterns rested directly on top of their facilities. Sellafield had never been accused of contaminating the local sheep populations previously, but government scientists' assurances of the temporary effects of the radiation exposure made the farmers suspicious. The swift and indeterminate ban on selling sheep and the coincidental localization of radioactivity around Sellafield led many farmers to accuse scientists of colluding against them. As Wynne (1992) notes, in "the manifest scientific confusion and inconsistency, it was as if the farmers had suddenly found an outlet for fears and suspicions that they had previously entertained" (p. 288). Skeptical of the scientists' assurances that Sellafield was innocent, they expressed extreme doubt in Wynne's interviews.

Wynne's quasi-ethnographic approach successfully demonstrates how attending to the lived ecology of subjects, and in particular their vernacular, can produce results with strong ecological validity. One of the most cited articles in *Public Understanding of Science*, Wynne's investigation of Cumbrian sheep farmers is frequently invoked by public engagement with science researchers as a canonical example. His analysis succeeds by delving into the region's history, geography, and industry, positioning stakeholders within that larger ecology, and taking seriously the farmer's skills, knowledge, and know-how of their trade. Carefully balancing the scientific facts of the case against the backdrop of a larger ecology of coincidences, spatial distributions, and arguments, Wynne's quasi-ethnographic approach gives a complex and compelling perspective on this peculiar set of circumstances. While the goals of organic public engagement differ from Wynne's study, his knowledge of the region, environment, and situation all demonstrate a commitment to more ecologically valid analysis.

Wynne does not interfere with, ignore, or discount the claims of the public under investigation. In fact, he goes to great lengths to show how these farmers attempted to leverage their practical know-how to aid scientists in conducting soil tests, but were ignored (Wynne 1992, p. 287). Walking the tight rope between over-valorizing vernacular discourses and automatically assuming scientific authority, Wynne produces a dynamic overview of this particular public ecology that demonstrates strong ecological validity.

Neither surveys nor media studies would be able to achieve this level of ecological sensitivity. While a survey study of farmer's attitudes toward the government's actions might have yielded important information, as a standalone methodology, it would not produce as equally ecologically valid data as Wynne's quasi-ethnographic approach. Unaccompanied by other metrics, surveys would fail to capture the context for the mistrust that farmer's felt toward scientists (a mistrust instilled because of the communicative failures and missteps of these scientists), making their antagonistic views toward science particularly discreditable given the forced limited choice of surveys, despite due justification for their skepticism. Alternatively, studies solely composed of popular press or media portrayals of the event would probably miss or dismiss the "conspiracy theories" of these local farmers, thus giving the impression that their claims (while technically incorrect) were not justifiable in the broader context of events. Wynne adeptly captures a coherent image of how this public actually lived in its ecology, and in doing so, struck a blow to reductive views of scientific literacy. Not merely a matter of "misunderstanding" the science, this particular public's lived experiences suggest they were reacting to a breach of trust between experts and nonexperts. Therefore, Wynne's quasi-ethnographic approach produced a more ecologically valid argument by capturing more of the connections, frustrations, and disruptions of his research site.

Another example is Stone's (2002) "Risk Perception Mapping" (RPM), a methodology utilized for examining the Great Lakes environmental management. RPM is a quasi-ethnographic approach that defines the geographical reach of an event's effect on a population, identifies the characteristics that comprise that population, and accesses and incorporates their knowledge base into decision-making. Different from Wynne's approach, which is more interested in vernacular and less interested in incorporating local knowledge into decision-making, Stone's methodology represents a sophisticated quasi-ethnographic approach to public engagement.

One reason is RPM's tendency to build up participation of a locally affected population, rather than having participants "self-select." In addition, Stone's ethnographic sensibility is complemented by a full range of other methods, including survey studies. The fieldwork alone involved contacting sample households and choosing respondents, administering structured interviews, resolving interview refusals, engaging in participant observations, conducting informal interviews, and reviewing media coverage and historic documents (Stone 2002, p. 100). Together, these varied and mixed approaches were far more likely to produce ecologically valid results, due to RPM's basis as an "emic" as opposed to "etic" approach.

Coined by anthropologist Kenneth Pike in the late 1960s, the terms "emic" and "etic" have been vital to differentiating methods of studying ethnographic participants. As Stone (2002) explains "an etically-conceived public derives from human organization definitions imposed from the outside, which may or may not have any basis in social reality" (p. 135). Alternatively, an "emically-conceived public…derives from human organizational definitions based on social interaction in cultural context" (Stone 2002, p. 135). Demographic criteria are examples of etically defined social groupings, because publics do not generally organize themselves into behavioral units that correspond to these criteria. Public engagement events have traditionally gathered participants based upon etic criteria. However, an emic methodology delineates publics as they behave in their ecologies, instead of imposing etic criteria from the outset.

Stone's (2002) dairy farmer anecdote is useful for explaining why this differentiation is important. Stone (2002) was an ethnographic field manager on an RPM study for a proposed radioactive waste facility when he encountered self-defined "milksheds," or "extensive collection zones for milk harvested by dairy farmers" at the research location (p. 136). Some of the members of these milksheds showed a high level of concern about this waste facility, even though they lived relatively far away from the proposed site. As Stone discovered, these dairy farmers had their milk mutually collected and processed by farms located near the proposed waste site. Their concerns were related to their membership in their respective milksheds, not to their profession as a dairy farmer, illustrating an emic as opposed to etic approach. Stone (2002) explains how without "this emic understanding subsequent participation programs would have homogenized the LAP [locally affected population] by presuming that its members' behavior was dictated by etically derived categories" (p. 136). In other words, the distinction between emic and etic is "a distinction between the entities and process of social life that are real and important to the participants versus entities and processes which by virtue of their scientific status are capable of efficaciously explaining (and changing) social thoughts and activities, regardless of whether they are real or important to the participants point of view" (Harris 1976, p. 330). Etic approaches artificially differentiate publics from the outset based on arbitrary criteria instead of developing a descriptive basis relating to actual behaviors of publics as they exist in their ecologies. Emic approaches are both under-utilized by public engagement with science researchers and necessary if strong ecological validity is desired in public engagement events.

Another quasi-ethnographic project that demonstrates strong ecological validity is Mol's (2002) investigation of atherosclerosis. Rather than focus intensively on a single facet of the hospital or organize her analysis around etic groupings, Mol's methodology hinged on examining multiple realities, or what she calls multiple ontologies, that make up the meaning and movement of atherosclerosis in a hospital. As Hine (2007) notes, this "means that her account becomes an account of locatedness in a deeper and more inquisitive sense than would be permitted by exploring any one of the forms of atherosclerosis that she encounters in isolation" (p. 664). During her study, Mol witnessed several operations; followed radiologists and pathologists; spent time in a medical lab observing hematologists; held interviews and conversations with epidemiologists, physiologists, internists, surgeons, and general practitioners; tracked down scholarly and historical texts; coauthored articles with doctors; supervised medical students who were interviewing vascular surgeons; and had a research assistant conduct and transcribe interviews with patients (Mol 2002, p. 3). In combination, this data will be more robust than if she simply remained in a single, constrained ecology. Much like Wynne and Stone, Mol's itinerant yet geographically constrained quasi-ethnography demonstrates stronger ecological validity because of the diverse steps she takes to capture the different aspects, facets, and faces of atherosclerosis.

What distinguishes Mol's methodology from other ethnographic practices, however, is her commitment to performing an "empirical philosophy." Rather than a field report, Mol's (2002) approach was to try to understand atherosclerosis as a material manipulated in practice. If, instead of "bracketing the practices in which objects are handled we foreground them," we can open up multiple realities (p. 4). As she explains, the "body, the patient, the disease, the doctor, the technician, the technology: all of these are more than one," however, "far from necessarily falling into fragments, multiple objects tend to hang together somehow" (Mol 2002, p. 5). Ecologically aware, Mol's conceptualization of her ethnographic methodology signals a shift from epistemological to ontological accounts of reality. In turn, this shift forces researchers to engage more deeply with their participants. An epistemological account is what underlies much of public engagement with science research. For instance, "the public's" engagement with science is thought to be assessable by tinkering with lived experiences and flattening out the complexity of how publics function; hence, artificial settings and environments are utilized without addressing questions of ecological validity. A shift toward multiple ontologies (a shift toward public ecologies) suggests that bracketing the richness of public participation and deliberation will not produce ecologically valid data. Reality is not merely fragmented as simplified readings of postmodernism would suggest; rather, reality is paradoxically both fragmented and tends "to hang together somehow," which leads us to believe that a quasi-ethnographic, vernacular, unfettered, and ecologically oriented public engagement methodology (an organic public engagement methodology) is invaluable in developing a better understanding of the complex behaviors of publics.

These examples provide different lessons for public engagement practitioners who wish to develop more ecologically valid research. Wynne's quasi-ethnographic observation of Cumbrian sheep farmers is useful because it demonstrates how the ethnographic sensibility is practical, valuable, and able to capture contexts inaccessible through most other means of researching public engagement with science. Alternatively, Stone's RPM is vital because it blends ethnographic methods alongside other qualitative and quantitative measures. In addition, Stone's approach maintains the interventionist goals of public engagement, but builds emic categories in a "bottom-up" fashion, instead of predetermining etic groupings in a "top-down" manner. Finally, Mol's multi-sited approach, a topic we will examine more in-depth in the next chapter, allows her to gain a clearer picture of how multiple publics encounter, frame, and understand atherosclerosis. Organic public engagement builds off Wynne, Stone, and Mol's methodologies, while retaining a distinct character that is tailored for public engagement with science researchers and practitioners.

What Is Organic Public Engagement?

Unlike traditional ethnographies, organic public engagement intervenes in the sites it encounters, develops broader studies of multiple sites rather than deep studies of few sites, and builds mid-level theories that can inform decision-making. The extent to which a researcher adopts one or more of these elements, they may be accused of doing less pure ethnography, but organic engagement walks the line between the deeply descriptive task of traditional ethnography and the goals of proactive public engagement and theory construction. Like any method, ethnography has limitations, but when one asks about discourses and practices within existing publics, an ethnographic sensibility strengthens ecological validity while avoiding the intentional fallacy. Some intervention into a public might be required to introduce a topic or to enable observation, but the guidelines for researchers concerned with ecological validity should be first, to disturb as little about that public and its ecology as is possible and, second, when changes must be made, they should mirror how they might organically occur within that ecology.

If the goal is to actually engage publics, not merely to observe them in their pristine state, we presume that engagement not only brings something to publics but, hopefully, also returns with something from those publics. Taking for example the research on public engagement with nanotechnology, organic public engagement almost certainly needs to introduce information about that topic into most publics. Very few individuals know much about nanotechnology and while groups are likely to have more knowledge (both accurate and inaccurate), if we want to study publics and how they might react to nanotechnologies, researchers will often have to introduce new information. Whether this diminishes the ecological validity of the research will depend upon how the new information is brought into the public being studied. Ecological validity is not an all-or-nothing standard but, like other forms of validity, an objective toward which researchers strive. The researchers introducing nanotechnology should understand the publics being studied well enough to know how such information might normally reach them. What are their means of bringing information into the ecology? What sources do they tend to rely upon? How is information disseminated within that ecology? With this kind of understanding, a researcher can bring new elements into an existing ecology and retain as much ecological validity as possible while still achieving the research objectives.

Alternatively, and perhaps ideally, researchers can reduce intervention by providing only an initial prompt and allowing an existing public to seek information and sources independently. Gehrke's (forthcoming) recent study of public perceptions of nanotechnology has confirmed Wynne's (1991) findings that "those who do have or develop motivation often show great alacrity at seeking out sources and assimilating science" (p. 118). Gehrke coordinated with eleven local groups across the United States, of which eight were willing and able to locate their own information, organize their events in conformity with their norms, and select and invite speakers without intervention by the researchers. One group selected not only an autodidactic approach but also to self-engage with core group organizers researching the topic and then presenting it to other regular members. By allowing each local public to take control of the setting, mode, and outcomes of the engagement, this method ensured as much as possible that information brought into these publics was guided by their norms, values, and existing behaviors.

This method also reduced costs and administrative burdens, enabling not only stronger ecological validity but also greater efficiency. As Maria Powell and Mathilde Colin (2009) have noted, public engagement events tend to be expensive and labor-intensive. While organic public engagement still requires resources and coordination, the capacity to utilize existing publics, their networks, and established communication channels can reduce some of those costs and burdens. Existing publics avoid some of the problems Powell and Colin (2009) identify as endemic to engagement with "unorganized lay citizens" (p. 328). Regardless, good organic public engagement research requires careful and potentially expensive analysis of detailed records of the events to produce insights that can be brought from publics into policy making as well as improve our understanding of how diverse publics perform citizenship and practice politics.

That second part of engagement, bringing something back from the publics being studied, should include two dimensions: analysis of observations and member-checks conducted with that public. In the course of an engagement event, observers should record the event in as robust a medium as the setting allows without significant interference. Accordingly, organic engagement can discover the "provisional rules" of vernacular rhetorics that "are created within the context of the dispute" (Sloop and Ono 1997, p. 56), as well as find what Michel de Certeau (2002) called the "figures," "turns," "verbal economy and condensation, double meanings and misinterpretations, displacements and alliterations, multiple uses of the same material" (p. 39), elements largely absent from previous public engagement research. Mapping out members' receptivity to information, redeployment of outside sources, frequent use of metaphors or analogies, demonstrations of value hierarchization, and so on provide guidance to decision-makers about public desires, expectations, risk tolerance, and values. Perhaps even more important, such mapping offers a pathway to understanding the various ways of speaking and reasoning in heterogeneous publics.

Such studies involve a double-caution reflecting the importance of an ethnographic disposition: organic engagement research should neither import a presumed model of good communication nor uncritically valorize local practices as without detriments. In the first case, we must be cautious to remember, as Hauser (1999) wrote, that rhetorical transactions set the foundations of publics: "to dismiss these transactions as flawed is to dismiss the means by which society posits the order of its own representations" (p. 273). Differing publics communicate and reason in diverse ways grounded in social, cultural, and institutional norms. This holds equally true for both nonexpert and expert communities. Wynne (1992) noted that the perspectives of scientists studying the risks and effects of the Chernobyl nuclear disaster were "just as socially grounded, conditional, and value-laden" as those of the farmers and community members with whom they were interacting (p. 297). Inversely, Ono and Sloop (1995) caution against idealizing vernacular discourses (p. 25). Thus, while we strive to avoid the judgmental model of deliberative democracy idealists, we likewise cannot fall prey to what Martyn Hammersley (1990) called the "urbane romanticism that celebrates the diverse forms of rationality, skill, and morality to be found among ordinary people" (p. 600). Durant (2011) accused Jasanoff and Wynne of precisely such romanticism and noted the potential costs of uncritically celebrating local or vernacular practices and discourses. Regardless of the merit of his specific accusations, the general caution against what Durant called "identity politics" cautions organic public engagement researchers to avoid unthinking adulation of the vernacular and the local.

Additionally, after the event and completion of the analysis, researchers should return to the public being studied and conduct member-checks. This minimizes observer mischaracterization of phenomena that occurred in the engagement event. It also makes the researcher responsible not only to the granting agency and decision-makers, but directly to the publics engaged. If a public finds the researcher has maligned them or misunderstood the dynamics and relationships in that group, they are often quite willing to say so. In this way, the second phase of engagement reduces the risk of an intentional fallacy caused by observer interpretation and gives publics greater control over the research outcomes.

Six Principles of Public Ecologies

When working with publics from an ecological foundation, we recommend keeping six key principles in mind. These six principles remind us that publics, as ecologies, are not merely aggregates of individual beliefs and behaviors, as so much of the social psychological research assumes. These principles also caution us against many of the excesses of conventional public engagement methods while reminding us of the unique value of the organic public engagement method. Keeping these six principles in the forefront throughout the research process, from initial design to final reporting, can make researchers more attentive to their practices and their full potential.

1. Publics as Ecologies Can Be Synergistic The most basic principle, and the one every student of group dynamics knows, is that any social organism develops synergies. A public, even if it is merely a dozen people, has capacities that significantly exceed the sum of the individuals who make up that public. Even for simple research on information deficits, publics will display knowledge and understanding that exceeds what any one of the individual members show, and exceeds the collective sum of their knowledge. For example, in Gehrke's research, a rural church group in a southern state had a conversation about nanoparticles that was not only surprisingly robust (at least a surprise to the researchers) but also showed them refining their understanding through their own in-group interaction. As they discussed how size of particles affects total surface area (and hence reactivity), the members questioned each other, provided analogies or illustrations, and had multiple moments of realization that brought every member to a richer relationship with the information.[2] For most of that conversation, the "expert" they had invited and who had just finished a presentation on applications of nanotechnologies was not involved in the conversation (and not invited to participate or given deference by the community members). At the same time, synergies are not always positive and publics as ecologies can also reflect negative synergies, where they are more capable of generating mistaken understandings.

A community advocacy group in the mid-west found themselves hooked on discussing scenarios of autonomous nano-robots going haywire and self-replicating beyond control, effectively destroying all other matters on the planet (known as the "gray goo" scenario). Such a scenario is only present in science fiction stories and the most fantastical of futuristic hyperbole, but in that public, for those fifty people, it could have been as real and present a threat in their lives as handgun violence or heart disease.

Letting those synergies express themselves, both positive and negative, is an important part of preserving the ecological validity of the research, which will be essential to understand why one public develops one type of discourse about the engagement topic, while another public has an entirely different conversation. Only by tracing those conversations, recognizing the synergistic moments (both positive and negative), and mapping how those synergies function can researchers come to say what causes something like a fear of "gray goo" to manifest in some publics but not in others. The resulting knowledge will be important not so much because it tells us how to dissuade a public from believing in such scenarios, but because it tells us the underlying values, ideologies, beliefs, and relationships that make the scenario compelling. Absent that knowledge, interventionist models of public engagement never meet actual publics.

2. Publics as Ecologies Can Be Self-Organizing Existing publics are already, in an informal way, organized and organizing. As such, they will usually have patterns or traditions of how they engage each other, how they interface with people or groups not identified as part of their public, how they acquire and process new information, and often who speaks in what roles and in what order. This self-organization may be codified in the bylaws of a civic association, or so ingrained in the culture of a public that the members have ceased to become aware of them, simply performing the norms without consideration or awareness. Regardless of the formality or transparency of their self-organization, this means that any given public already has methods of engaging a topic, acquiring information, and deliberating.

If our goal is to understand publics, and we hold that publics are ecologies (even if also existing within larger ecologies), then the best modes of engagement are those that let the publics choose and organize the engagement. In a study engaging eleven publics across the United States, Gehrke found eight of these eleven publics strongly preferred to organize their own engagement events, handling every aspect of the event on their own. This included site selection and arrangements, choice of format, inviting perceived experts (or not), choosing the format (open discussion, presentation with a question period, a debate, a conversation over dinner, etc.), and even producing and distributing their own written materials. The remaining three publics preferred slightly more direction or involvement from the researchers, but in no case did any public require the researcher organize the entire event. In fact, in every case the publics articulated either that they did not want the engagement event to disrupt their normal modes of interaction, or that they wanted to control the theme of the event to conform to their own collective identity.

Formal civic associations, regardless of location, tended to be concerned both about having their normal meeting processes disrupted and about being exploited for commercial or ideological purposes. A similar concern was likewise expressed by a much less formal local "slow food" group. Meanwhile, an environmental group in the western United States had no concerns about format (though they did decide to invent their own), but were deeply concerned about the potential for such events to reflect a pro-industry bias and wanted to control who was invited to speak.

Respecting the self-organizing capacities of publics not only strengthened the ecological validity of the research, allowing researchers to see how publics actually choose to engage on such topics, but also made the research dramatically more efficient and easier to organize. By cooperating with an existing public, which often has its own access to physical space and means of publicizing events, and encouraging them to handle all the logistics, researchers can focus on data collection and analysis. The eleven events in Gehrke's study cost, on average, less than $1,000 each to run while taking less time and energy from the research team than conventional public engagement. As a result, the vast majority of the grant funding went to improving the quality of the data collection and analysis.

3. Publics as Ecologies Can Vary in Resiliency A public continues to exist as a public because of a set of relations and way of cooperating exists between the members, and because there is a collective commitment to maintain the group relation or identity, even if the particulars of how they organize and cooperate might change. The strength of that commitment is one of the variables that will affect a public's resiliency. A certain tension may exist between the strength of that collective identification or bond and the formality and rigidity of a public's processes of self-organizing. In times of change or challenge, a public may be confronted with changing its modes of relating, its collective identity, its membership, its geographic location, or perhaps all of these. At heart, the value of association (which may be deeply political, simply social, or even merely convenience) holds the public together, serving as a force of attraction and as a membrane around that public. Absent a sufficient energy of association (which can also be shared misery or simple loneliness), a public will reorganize into new publics, be absorbed into other publics, or dissipate.

Resiliency of a public can be difficult to study, even in more formally organized publics. A local chapter of a national civic association may, at first, seem quite resilient and durable. Perhaps it has existed for decades, through multiple changes in membership and mission. However, the continuity of the formal association should not be equivocated with the resiliency of the public that might emerge within that formal association. Even if a club, chapter, or organization continues to exist, the public inhabiting that formal shell can be displaced, sometimes completely reversed or radically reconstituted, perhaps over a slow change in membership and norms or perhaps in a dramatic shift in identity. Student organizations on college campuses sometimes manifest these kinds of transient and fragile publics that encounter major shifts in culture, form, constitution, and purpose in relatively short periods of time, even while inhabiting durable formal associations. Such formal associations can be like the shells of hermit crabs, being periodically vacated by one public and then inhabited by another, sometimes passed on multiple times over long periods.

Inversely, sometimes even very loosely organized and informal publics that appear to have no durable name or structure demonstrate significant resiliency over long periods of time. These can be especially difficult to identify or trace, precisely because the more traditional and visible markers of association are less prominent, but may also be more revealing about how publics form and shift. For example, the local and slow food group in Gehrke's study would be difficult to organize in a formal way and had no discernible structure or organization beyond some loose sense of affiliation, a common investment in local or slow food, and some general practices of decorum that kept inviting the same people together and allowed robust discussion without threatening their collective hospitality. Farmers, ranchers, foodies, chefs, and restaurant owners, often with significant differences in political philosophies, nonetheless sustained their affiliation.

Giving due attention to resiliency can be difficult in public engagement research, especially since a researcher may only encounter a public for a short period of time (usually less than two years, often less than one). At minimum, the principle of varying resiliency reminds us that the public we are studying today may be able to withstand (or even absorb) disruptions, or may be fragile and shatter with only minimal outside force. In studying publics, their resiliency or fragility may be masked by the presence of a formal structure that serves as a wrapper or shell, while their substance is completely replaced or even withers. When researchers have the opportunity to witness publics going through transformations or confronting challenges, understanding their degree of resiliency and how that resiliency manifests or functions yields a richer and more useful understanding of not only that public, but other publics as well.

4. Publics as Ecologies Vary in Flexibility Just as the self-organizing principles of publics take on varying degrees of formality, so too do they vary in their flexibility. Some publics maintain their resiliency and direct their change through strict rules, determined procedures or processes, and even codified bylaws. Others achieve those goals with much looser norms that flex from moment to moment as circumstance changes. That flexibility can shape what is possible for a public engagement researcher, and also dictate how careful the researchers must be to avoid disrupting that public as an ecology. A flexible and resilient public may be much more capable of handling a disruptive mode of public engagement while retaining and displaying their own norms as a public and modes of discourse. However, an inflexible public, even if particularly resilient, might permit a highly disruptive mode of engagement, and then operate wholly differently in that engagement event than it does through its own norms.

The real challenge is that the flexibility of a public as an ecology is less about what they are willing to do than about at what point a specific public ceases performing its status as that public and becomes something else. An inflexible public may acquiesce to researchers' requests for format or content (especially when incentivized) but behave, communicate, and even become an entirely different organism in the research setting. Compounding this challenge, researchers will likely find it impossible to determine the flexibility of a public prior to engagement. The problem of response bias makes it infeasible to simply inquire of a public its flexibility. The only real solution for an organic public engagement researcher is to direct each public as little as possible, allowing its relative flexibility to direct its own choices. Not only should researchers not seek to guide publics in their format, scope, or method of engagement, but researchers must also actively resist giving guidance to publics even when they ask for it. When advice must be proffered, the optimal method is interrogative; the researchers should inquire about recent events or activities that public has undertaken, about their normal processes, and about things they decided not to do and why. While response bias is still a concern in such conversations, it may at least be mitigated by not directing the public toward a solution familiar to the researcher and alien to the public.

Paying attention to the flexibility of publics not only helps to preserve the ecological validity of the research by diminishing researcher intervention, it also allows both the static traditions and robust inventional capacities of publics to guide future engagement events. In Gehrke's study, a minority of the publics engaged had well-established modes of public engagement that engagement researchers often view as overly simplistic and less efficacious. Four of the eleven publics studied had only one model for engagement: the lecture followed by a question-and-answer period. While very few engagement researchers consider this an ideal mode of engagement, to introduce any other method would have been a significant disruption to the communicative and deliberative norms of these publics. Interestingly, in observing these events, Gehrke's research was able to map some of the dominant lines of trust in these publics, where they seek information, and some of the metaphors and narratives that drove their biases against seemingly benevolent technologies (such as cancer treatments). We contend that by respecting their inflexibility and their local traditions, following their lead in the content and structure, researchers are much more likely to see public perform and communicate these kinds of practices and biases.

5. Publics as Ecologies Can Evolve The evolution of a public is not necessarily its improvement toward a moral, political, or theoretical perfection. Instead, by evolution we mean something much closer to how biologists use the term: diversification and proliferation through apparently random mutation or hybridization, with the potential to manifest in greater fitness of the species to its environment. Evolution in publics can be accomplished through synergies, both positive and negative, as a result of flexibility (or inflexibility) and resiliency (or even the lack thereof).

Every public is both an ecology and at the same time part of a larger ecology, just as is the case with ecologies in biology, toxicology, or geography. Ecologies overlap, interact, subsume one another, and sometimes even assimilate or strangle other ecologies out of existence.

When we speak of evolution (whether it be of publics or species), we should divorce this principle from any valuation, any sense of the good or the just, and accept that the changes in organisms (publics or viruses) are less an effort to approximate perfection than a result of the imperfection of replication and self-perpetuation. Evolution is the expression of the imperfection of an organism's ability to replicate itself. All organisms, social, political, and biological, risk mutation, at least a little, every time they replicate. Often in seemingly random ways, each replication may introduce a variation, most of which will be completely invisible or even insignificant, but some of which may be dramatic shifts in the ability of that organism to survive or continue replicating.

Evolution is often easy to observe in viruses, not because they are somehow better at adapting or because their changes are more determined by their environment. In fact, there is no evidence that the evolution of viruses is any "smarter" or more directed than the evolution of any other organism. The apparent success of much virus mutation is a result of the size of their population, their speed of replication, and the frequency with which they mutate. The influenza virus (the common flu) replicates rapidly, easily infecting an entire office building within a matter of hours, reaching a population orders of magnitude larger than humans have ever seen, and with a greater chance of significant mutation in each replication. The flu virus evolves rapidly (and hence defeats our efforts at a durable vaccine) simply by relying on the law of large numbers. A trillion replications with a high chance of variation on each replication has a good chance of creating the one-in-a-billion random mutation that allows the virus to bypass the vaccine. Essentially, if you think of evolution like playing random lottery numbers, the flu virus buys vastly more tickets more frequently than the human species, and has more variation in the numbers on its tickets. Hence, its odds of hitting a genetic "jackpot" are much higher.

While we find it tempting to think of publics as having more directed or deliberate practices of "evolving," we believe those kinds of directed and deliberate activities might better be considered a part of their flexibility and resiliency. Preserving the principle of evolution for the inevitable and persistent random mutations and hybridizations that happen in the regular course of replication allows us to study shifts in publics as ecologies interacting with other ecologies, without the assumptions of intent or direction.

6. Publics as Ecologies Can Vary in Porousness Some publics seem to be almost seamless, moving in fluid ways, constantly taking in new members and information while also flowing out into adjacent and surrounding ecologies. On the other hand, some publics seem nearly impermeable and shut off from the world, perhaps by formal restrictions and rigid policies but just as likely due to informal norms of the culture. In addition to the degree of porousness, publics vary in how they control the inflows and outflows, what kinds of elements they allow to come into the public and what they allow to flow out.

Individual gatekeepers and collective filters are two ways that publics control their porousness, and both can be observed by researchers employing an ethnographic sensibility to observe the behaviors and discourses of a public. For example, some publics display a strong local bias, not only limiting their membership to a small area but also demonstrating higher trust and greater openness to local sources of information. Gehrke found this to be especially true of publics that organized as local chapters of national or international organizations. Regardless of the nature of the organization, their localism manifested in both a greater focus on how a topic or issue impacted their local areas and a strong preference for local or at least regional sources of information. Local chapters of Lions or Rotary were more likely to seek information from a local source and much more likely to discuss research, industry, or policy in their own county than publics that were not subgroups of a larger organization. Even local consumer advocate associations, church groups, and student organizations tended to have a greater openness and curiosity about sources of information and issues that were geographically distant. The one exception to this tendency in his study was the slow/local food group, whose localism was deeply engrained in their culture.

Porousness may also manifest in how the public extends itself outward, disseminating information or even having sub-groups of that public engage with other ecologies as representatives of that public or perhaps as marked dissidents. Such porousness may generate splinter groups, new publics, or hybridization in other existing publics, creating a new ecology and a new public in the process. This kind of emergence may function like the tributaries of two rivers merging to form a third river, like invasive plants transforming a landscape, like one subspecies mating with another producing something new, or in any number of other patterns. Such transformations may themselves be the result of other shifts in a public ecology or in one or more of the surrounding or adjacent ecologies interacting with that public. Hence, resilience and adaptability may interact with porousness to affect the evolution of a public.

In any case, researchers can learn much by paying attention to the porousness of a public ecology, both its degree and its kind. Doing so may reveal why a public can remain recalcitrant in certain beliefs or practices, even while surrounding public ecologies have shunned both those beliefs and the people who hold them. Studying the porousness of a public may also teach us the pathways that can be used to build interactions between publics who otherwise might be closed to each other, either procedurally or discursively. Without the ethnographic sensibility and the low-intervention preservation of ecological validity, researchers are highly unlikely to locate and reveal these qualities in publics. Hence, even those desiring to make an intervention will find themselves stymied by their incapacity to understand how a given public could possibly retain a certain viewpoint, a set of beliefs, a value hierarchy, or a body of practices. Ironically, it seems that even high-intervention cynical and idealist engagement practices can only achieve their goals if they have the kind of knowledge that can only come from a low-intervention mode of engagement, such as the organic public engagement methodology.

USING ORGANIC PUBLIC ENGAGEMENT

Before moving much further into the mechanics of the organic public engagement methodology, we do want to caution against the possibility that researchers will deploy the method in decidedly inorganic ways. An organic approach to public engagement differs from conventional approaches not only in its low-intervention methods and observational approach, but most importantly in its attentiveness to the local conditions of the public being studied. In fact, the low-intervention methods and quasi-ethnographic methods are expressions of a core principle of organic public engagement: adaptation of method to the existing local conditions of the public being studied. Just as organic farming is, ideally, not simply about avoiding chemical fertilizers and pesticides, but also works with indigenous plants, the local soil and water, and similar conditions to adapt the farming to the given ecology, so too should organic public engagement work with the local conditions and adapt its methods.

Hence, what we offer here is not a method that researchers should simply apply uniformly to any given public, but more a blueprint or recipe for a method, or the logic of a method (a methodology) that a researcher must tinker with and modify each time it is deployed, so that it fits the public(s) being studied. Just as generic plans for a house must be modified to the building site, a research method must be modified to each given implementation. There can never be one off-the-shelf model of organic public engagement, but only prototypes or specific manifestations that represent a tinkering with local knowledge and found resources. Each method, including the one we are describing here, is hence always provisional and no hard rules can exist.

We have already said a great deal about the general logic and principles of organic public engagement, so rather than repeat those themes here, we offer an example of how one organic public engagement study was conducted. Again, this description of the method used in this study is meant to be illustrative and spur the imagination and invention of researchers, not draw a strict boundary around the method. We have based this description on Gehrke's (forthcoming) four-year study of public perceptions of nanotechnology across the United States.

As usual, the researcher began by defining some research questions. Initially in the research design these included what publics believe about the risks of nanotechnologies and why they held those beliefs. As the design matured, the questions shifted to ask how qualities of differing publics might influence their formation of opinions about nanotechnologies. This was already an adaptation to the found publics, the vast majority of which displayed neither significant interest in nanotechnology nor any real opinions on the topic. In the researcher's view, the question of what publics believe and why was premature. Either one would first have to ask what publics are taking an interest in nanotechnology and why, or one would need to ask the hypothetical questions of what a public might believe if they encountered the technology and how would they form those beliefs. Either is a reasonable approach to the research, but the researcher chose the latter due to the role the public engagement played in a larger research grant. Researchers and research projects are themselves also always playing parts within their own ecologies.

Given that the topic of inquiry was public perceptions of nanotechnology and practical consideration necessitated choosing a small set of publics for study, the researcher began by asking where in the United States existing interest in nanotechnology was high or might have a greater chance of becoming more prominent. To answer this question, the study relied upon previous research that had mapped nanotechnology-related industry, research, and other activities.[3] In the interest of observing a variety of publics, the possible locations provided by the maps were divided both by the type of nanotechnology work being done (scholarly, industry, governmental, etc.) and the type of locale (rural, suburban, and urban). Possible sites were then distributed across the United States to ensure some representation from the West, Mid-West, Mid-Atlantic, and Southern states. Because resources allowed for less than a dozen total public engagement events, not every possible permutation was included in the study, but the researcher did strive for a diversity of sites.

Members of the research team then researched the sites to find publics. This was the most difficult part of the engagement process, both because locating local publics from across the country poses challenges and because local publics are often not keen to coordinate with researchers with no connection to that public or its locale. This was mitigated to some extent by relying on local personal and professional connections (including not only other researchers in those locations but in some cases family members or friends). The method of initial contact was guided both by what could be gleaned from the norms of the public being contacted and by the pathway that had led the research team to that public. In some cases this was a phone call, in others an email or a letter by post. In a few it was an informal conversation via a local proxy.

Not all publics were interested or willing to participate in the research. Some expressed no interest in the topic, some displayed closed ecologies and were not open to outside interaction, and some simply said they were busy with other projects. While we think it reasonable to be concerned that such research can miss important publics, particularly if their nonparticipation is due to low porousness and inflexibility, the process also demonstrated that some publics are far more likely to take up the topic than others. In any case, the result was eleven differing publics across the United States participating in the research.

As each public joined the study, the researchers did as little as possible to organize, plan, or guide their engagement events. Instead, they were asked about past events they held, their normal format, the kinds of things they like to do or would like to be doing, and then asked to organize and hold the event themselves. The research team did provide modest financial support, from a National Science Foundation grant, to each public to help defray their costs. The largest commitment was less than $1000, with most being closer to $500. Throughout this process the research team strove to be hands-off and not intervene in the local public's choices. Two groups did ask the research team to help locate speakers, which was done by asking the representatives of that local public about past speakers and the kinds of speakers they preferred, then offering a few choices that met their criteria. Researchers did not ask to approve the speakers, for input on the event content or structure, or in any other way seek to influence what kind of event each public developed. The only thing the research team asked is that they be allowed to record the event so that it could be later analyzed.

In this particular study, the researchers were also using a pre-event survey instrument to build a snapshot of the demographics and knowledge of each public. The survey was done on paper, online, or both, depending on the preferences of the public group. To minimize framing or priming effects, the only questions about nanotechnology on the survey were two open-ended questions that asked what respondents knew about the topic. Such a survey instrument is not essential to organic public engagement and turned out to be of limited explanatory value. Actual observed behavior in the engagement events provided much richer data than the survey.

At each event the research team attempted to minimize their presence and involvement. In almost all cases, only one member of the research team was present at the event and that member had no role in introducing or managing the event. In no way did any member of the research team seek to interrupt, direct, or correct a speaker or discussion. In most cases, the member of the research team was silent for most of the event, perhaps only being introduced by the event coordinator at some point and briefly thanking the public group for holding the event.

A single high-definition video camera and microphone were set up in an innocuous location, set so as much of the total event was in frame as possible, turned on before the event, and then left unattended throughout the event. As is common in much research, a tension exists between capturing robust data and interfering with the research conditions. Use of a video camera may be disruptive to some contexts, though in this study it appeared that after a few minutes most of the public groups forgot about the presence of the camera. Researchers were mindful of the risks of video recording and worked with the local organizers to minimize disruption. It may be possible that field notes rather than video recording would be less disruptive and ensure even stronger ecological validity.

The use of video recording did allow a very detailed analysis of the events, including participants' nonverbal and paralinguistic behaviors. Analysis was done using open coding. Two of the primary investigators watched all eleven videos and made a list of topics, themes, metaphors and analogies, notable nonverbal and paralinguistic behaviors, and similar elements of the engagement events. Some of these were then assigned valences of positive/negative to provide further detail to the analysis. After merging their lists, two doctoral students in rhetoric were trained as coders and invited to offer additions, amendments, or edits to the list as part of the training process. By the end of the coder training, a final list of codes had emerged from the videos of the events and these were the set used in the data analysis.

The researchers and coders both agreed that some of the codes could only be used if they were coding the videos and not transcripts, so rather than produce text from the events for coding, the coders were trained to use the NVivo software for computer-assisted qualitative data analysis (CAQDA).[4] NVivo allows the direct importation and coding of audio-visual files in a drag-and-drop interface. Coders and some of the research team required training on the NVivo software, which has a significant learning curve. Once the training was complete, all of the codes were programmed into NVivo and the videos of the events were imported and ready for analysis. To ensure the ability to later assess inter-coder reliability, coders were instructed to code using full one-minute blocks of time (e.g. from 21.00–21.59). Thus, if audience activity was being coded for laughter, the coder would select the full minute in which the laughter began through the full minute in which the laughter ended and code it all as laughter. For example, laughter that started at 17:55 and ended at 18:20 would be coded from 17:00 to 18:59. While the set time blocks could be set to a smaller range (say, thirty seconds), without using standardized time blocks it becomes exceedingly difficult to leverage the power of CAQDA for coding and later analysis.

Each coder worked independently, coding every video by making two passes through the video in the NVivo software. When they were done, both coders' results were combined in the primary investigator's NVivo project file and compared using the Cohen's Kappa coefficient function in NVivo. This resulted in a list of codes with high inter-coder reliability. The primary investigator could then work from the cumulative data of the coders, finding common code frequencies and relationships, and also easily view all those segments of the eleven different events that were assigned the same code or set of codes. This allowed deeper qualitative analysis focused on a specific phenomenon or behavior. For example, the coders found a common trend toward negative audience reactions to nanotechnology's promise of cancer treatment. To understand why, the primary investigator watched those elements of videos coded in this way and discovered the negativity was due to audience skepticism fueled by the sometimes personal (and often lengthy) experience of being told that the next big cancer breakthrough is on the horizon. Similar nuance was found in a number of commonly occurring and highly reliable codes, such as audience distrust of government combined with audience expectation that government is the best source to provide protections for consumers.

This method provided a robust body of knowledge about how and why a variety of publics currently or will perceive nanotechnologies. Using an organic public engagement method, with its quasi-ethnographic approach and ecological orientation, the data retained high ecological validity and was robust enough to draw insights as to why and how publics form or express certain opinions. The use of CAQDA for data analysis and a well-trained set of coders produced a compelling combination of empirical and analytic results that helped reinforce the conclusions that could only be drawn through more nuanced qualitative analysis. The initial whitepaper produced by the researchers ended with its insights about public perceptions of nanotechnology, stopping short of developing theories or broader recommendations.

In later publication of the study, Gehrke extended the conclusions of the original whitepaper and used the data from the study to produce broader insights about public understanding of science, effective science communication, and the possibilities for responsive nanotechnology regulation. The next chapter provides both the principles for how those broader theories might be built from organic public engagement events and practical advice for researchers trying to build such theories.

NOTES

1. For instance, see Phil Macnaghten, Matthew Kearns, Brian Wynne, "Nanotechnology, governance and public deliberation: What role for the social sciences?" *Science Communication* 27, no. 2 (2005): 268–291.
2. For details on the study discussed in this section see Pat J. Gehrke, *Nano-Publics: Communicating Nanotechnology Applications, Risks, & Regulations*, New York: Palgrave, *forthcoming*.
3. For a current example of such research, see the Nanotechnology Project's interactive map (http://www.nanotechproject.org/inventories/map/).
4. NVivo is a product of QSR International. QSR was in no way affiliated with this study. Neither the researchers nor the authors have any financial interest in QSR International.

REFERENCES

Cicourel, Aaron. 2007. A Personal Retrospective View of Ecological Validity. *Text and Talk* 27 (5–6): 735–752.

Conquergood, Dwight. 1992. Ethnography, Rhetoric, and Performance. *The Quarterly Journal of Speech* 78 (1): 80–97.

de Certeau, Michel. 2002. *The Practice of Everyday Life*. Berkeley: University of California Press.

Durant, Darrin. 2011. Models of Democracy in Social Studies of Science. *Social Studies of Science* 41 (5): 691–714.

Gehrke, Pat J. forthcoming. *Nano-Publics: Communicating Nanotechnology Applications, Risks, & Regulations*. New York: Palgrave.

Hammersley, Martyn. 1990. What's Wrong with Ethnography? The Myth of Theoretical Description. *Sociology* 24 (4): 597–615.

Harris, Marvin. 1976. History and Significance of the Emic/Etic Distinction. *Annual Review of Anthropology* 5: 329–350.

Hauser, Gerald. 1999. *Vernacular Voices: The Rhetoric of Publics and Public Spheres*. Columbia: University of South Carolina Press.

Hine, Christine. 2007. Multi-Sited Ethnography as a Middle Range Methodology for Contemporary STS. *Science, Technology, & Human Values* 32 (6): 652–671.

Macnaghten, Phil, Matthew Kearns, and Brian Wynne. 2005. Nanotechnology, Governance and Public Deliberation: What Role for the Social Sciences? *Science Communication* 27 (2): 268–291.

Mol, Annemarie. 2002. *The Body Multiple: Ontology in Medical Practice*. Durham: Duke University Press.

Ono, Kent, and John Sloop. 1995. The Critique of Vernacular Discourse. *Communication Monographs* 62 (1): 19–46.

Powell, Maria C., and Mathilde Colin. 2009. Participatory Paradoxes: Facilitating Citizen Engagement in Science and Technology from the Top-Down? *Bulletin of Science, Technology, and Society* 29 (4): 325–342.

Sloop, John, and Kent Ono. 1997. Out-Law Discourse: The Critical Politics of Material Judgment. *Philosophy and Rhetoric* 30 (1): 51–69.

Stone, John V. 2002. *Public Participation in Environmental Management: Seeking Participatory Equity Through Ethnographic Inquiry*. PhD Dissertation, University of South Florida.

Wynne, Brian. 1991. Knowledges in Context. *Science, Technology & Human Values* 16 (1): 111–121.

———. 1992. Misunderstood Misunderstanding: Social Identities and Public Uptake of Science. *Public Understanding of Science* 1 (3): 283–304.

CHAPTER 7

Generating Outcomes from Organic Engagement Events

This chapter explores how organic public engagement is informed by several middle-range methodological approaches. Drawing from socio-logical research, we argue that studying public engagement events inspired by grounded theory and multi-sited ethnography produces more ecologically valid research. Grounded theory provides the ability to develop theories for understudied phenomena and draw new conclusions on existing studies. Many public engagements with science studies involve engaging publics with cutting-edge and novel scientific and technological developments, so being able to develop new and ecologically valid theories is imperative. Alternatively, many of the conclusions currently driving public engagement with science research struggle to break free from weak theoretical assumptions, like the deficit theory, so novel approaches to understanding existing interactions between publics and emerging scientific developments should be particularly appealing to public engagement with science practitioners and researchers. Multi-sited ethnography increases the applicability and validity of organic public engagement. A central issue for public engagement with science research is the possibility that by studying only single sites of engagement, researchers can become inculcated or develop distorted views of how publics engage with science. One way to avoid unfairly extrapolating the data gathered from a single public to "the public" is to study multiple sites where different publics engage the same set of scientific topics.

A.S. Lerner, P.J. Gehrke, *Organic Public Engagement*,
https://doi.org/10.1007/978-3-319-64397-7_7

Both of these approaches to theory-building are "middle-range" in nature. Discussed at length by sociologist Robert K. Merton, middle-range theory breaks out of the dual traps of ineffectual hyper-contextualized case studies on the one hand and chasing after ephemeral grand theories on the other.

MIDDLE-RANGE THEORIES OF PUBLICS

Sociologist Robert Merton (1968) described theories of the middle range as "theories that lie between the minor but necessary working hypotheses…during day-to-day research and the all-inclusive systematic efforts to develop a unified theory that will explain all the observed uniformities social behavior, social organization and social change" (p. 39). Primarily designed to guide empirical inquiry in sociological studies, these intermediate theories involve certain amounts of abstraction, but are close enough to observed data that they can be "empirically" tested (Merton 1968 p. 39). According to Merton, in sociological studies, a middle-range approach is developed by considering the elements of social structures, as opposed to detailing historical descriptions of individual social systems. For Merton, the middle range is all about finding a balance between historicizing and generalizing. The latter tends toward particularity, thus limiting the capacity of research to be applicable in other cases. The former tends toward universalization, thus limiting the capacity of research to pick up on the granular differences between various sociological observations.

When Merton first proposed a middle-range sociological theory, he was confronted by two major objections. On the one hand, sociologists engaged with theoretically oriented empirical research easily capitulated to Merton's suggestion, since in practice, middle-range theory-building had already been the working philosophy of these practitioners. On the other hand, those committed to the history of social thought or those who attempted to develop a "total" sociological theory "described the policy as a retreat from properly high aspirations" (Merton 1968, p. 54). In the middle were those who realized that middle-range theory does not mean sole attention to middle-range theory. Instead, middle-range approaches can contribute to developing comprehensive theories by consolidating empirical research rather than beginning with a top-down, "grand" theory approach, which often artificially constrains empirical data to preconceived categories.

Applied to STS, middle-range theories have been valuable for researchers wishing to bridge the gap between individual case studies and grand theoretical systems. In a special issue of *Science, Technology, & Human Values* on middle-range theory, Sally Wyatt and Brian Balmer (2007) argue that researchers in STS often run up against a widening gap between pure description and overgeneralized theory. Purely descriptive case studies offer limited value to producing connections between concepts in disparate STS scholarship, while grand theories like actor-network theory (ANT), social construction of technology (SCOT), or Collins and Evans' programmatic "Third Wave" are readily over-appropriated, resulting in claims that are neither novel nor particularly impactful. Wyatt and Balmer (2007) argue that a failure to develop middle-range theories not only prevents the development of credible and important theories within STS but also prevents STS from being able to engage a wider audience. As Wyatt and Balmer (2007) note, for Merton "middle-range theory meant engaging with reality...[and] producing theoretical accounts that engaged with that reality which themselves could be used to communicate with others, whether policy makers or scholars from other disciplines" (p. 621). Middle-range theory, then, is in alignment with our effort to construct public engagement practices that simulate the reality of publics' actual deliberative behaviors. One upside of this calibration to real situations is the potential to increase the impact, applicability, and translatability of data gathered at public engagement events. Other benefits include a space to reflect on the shortcomings of research designs or build bridges between cases, theories, and public policy.

For instance, Anne Beaulieu et al. (2007) argue that by exploring and middle-ranging ethnographic successes, failures, and shortcomings, we stand to gain a capacity to "makes sense of these successes and failures" (p. 673). Diverging rather significantly from Merton's formulation, Beaulieu et al. (2007) note how "the middle range is not a space of unification but the tracing of tensions and interrogations in our work" (p. 680). Like many studies that claim to seek a middle-range theory, Beaulieu et al. use the basic conflict between the particular and the general to generate practical self-reflection on their work. For example, they suggest that putting women's studies and e-science in conversation is useful comparative work for generating critical notions of diversity, inclusion, and knowledge-making in e-science. However, a comparative approach will "only show meaningful differences once cases are made comparable," and a "unit for comparison has to be constituted, and features for comparison have to be specified, if this approach is to yield interesting insights" (Beaulieu et al. 2007, p. 677).

Bridging approaches, theories, or case studies requires some common unit of measurement, therefore a middle-range approach confronts what counts as a reasonable comparison, what it means to be comparable, and how it is possible to know what constitutes a reasonable comparison. A middle-range comparative theory is built out of the case studies under comparison, instead of conjuring forth a generalized theory from the aether or from macro-analysis of previous research. Alternatively, a middle-range comparative theory denies that individual contexts are so rarefied that general principles are inextricable from research findings. Doing middle-range theory means asking questions that land neither in the complex particularity of the case at hand, nor the vague amorphousness of a unifying theory. Instead, middle-range theory addresses questions of what it means to perform comparative work in the first place or on what basis comparative work can be understandable, sensible, or coherent.

Beaulieu et al. demonstrate how middle-range reflection can help build plausible theoretical tools from individual case studies; however, middle-range theories are also valuable in bridging the gap between STS research and public policy. Frank Geels (2007) notes how while STS has developed as a vibrant and dynamic academic field, it often fails to translate to policy or other disciplines (p. 630). Geels's diagnosis of the problem is twofold: first, the proliferation of concepts and jargon has made STS research difficult to understand for outsiders. Second, an emphasis on complexity, contingency, and context can make it difficult to communicate general lessons from research. Geels argues that both problems are mitigated by taking a middle-range approach.

When applied to the problems of jargon and conceptual proliferation, middle-range theories attempt to pierce the theoretical fog that sometimes accompanies individual case studies. For instance, introducing a case and suggesting a new concept to assess the inner workings of a research site often leads to increasingly untidy, esoteric, and isolated theory-building. A middle-range approach alleviates this problem by pushing researchers to compare new concepts to existing clusters or groups of related concepts. Alternatively, detailed examinations of the complexity and messiness of interactions between actants, as in actor-network theory, were a successful and useful approach to STS in the 1980s and 1990s, but are "now suffering from diminishing returns" (Geels 2007, p. 631). As Geels (2007) explains, attending to this messiness set STS down both a difficult and repetitive path, since "science" and "society" will continue to make "wrong" claims about science and technology, making STS's task never-ending (p. 631).

At a certain point, claims about complexity, contingency, messiness, and situatedness simply lack novelty. Middle-range theories alleviate this problem by attempting to recognize patterns, styles, and regularities across situated contexts, thus preserving the inherent "messiness" of science while maintaining that local practices can be generalized across multiple situations.

When it comes to STS's inability to influence public policy, some of the problems of ineffectual public engagement no doubt rest in the inability to articulate a middle-range theory of publics. Without a realistic account of how publics actually respond to scientific information, public engagement with science researchers have found themselves without access to middle-range theories. That is, the "public" in "public engagement with science" tends to either be hyper-contextualized to the point where research findings are difficult to translate into other settings, or over-theorized, resulting in efforts to constrain public engagement to idealist frameworks of democratic deliberation. The former tends to find shelter in esoteric divinations of observed behaviors, thus making their results incoherent to researchers wishing to replicate public engagement methodologies for diverse publics. The latter tends to shield itself with a general vagueness, only surmountable by artificially manipulating publics' behaviors in order to fit a theoretical apparatus. The former breaks down the possibility of cooperative recognition of trends, patterns, and resonances across diverse publics. The latter breaks down the possibility of achieving ecologically valid research, since public engagement with science practitioners funnel themselves (and publics) into fixed theoretical passageways. Merton's conception of a middle-range, then, is a promising way to approach public engagement with science research, since it avoids many of the problems that befall social scientific and humanistic research.

What then does a middle-range theory of publics look like, and how can it be applied to public engagement with science settings? A middle-range theory of publics can be constructed by examining multiple publics in their respective ecologies. The seven principles of public ecologies described in the previous chapter are one such example of middle-range theory-building.

Organic public engagement draws on a middle-range theoretical approach, thus producing more realistic representations of how publics actually deliberate on scientific topics. As Wyatt and Balmer rightly argue, middle-range theories ultimately do a better job of accurately describing the reality of a situation, and this principle extends to public engagement events. Current attempts at public engagement neither allow publics to behave "in the wild" nor "engage" with them in an ecologically attuned manner. One way to prevent these problems from distorting public engagement data is to strive to utilize methods that can contribute to middle-range theory-building. After doing so, public engagement with science practitioners can work toward building more sophisticated, more realistic, more translatable, and more ecologically valid theories of how publics engage with science. We used two strategies to accomplish these middle-range goals: grounded theory and multi-sited ethnography.

GROUNDED THEORY

Sociologists Barney Glaser and Anselm Strauss coined their middle-range methodology "grounded theory" in the late 1960s. Since theory must fill many explanatory and predictive roles to be useful, Glaser and Strauss (1967) argue that the best approach is a "systematic discovery of the theory from the data of social research" (p. 3). The premise is simple: theory that best fits the data will emerge from the data. Much like Merton, Glaser and Strauss see "grand" theory as limiting the capacity of researchers to produce valuable, translatable, and sensible theories. However, unlike Merton, Glaser and Strauss systematize a middle-range theory-building program into a distinct methodology. In the late 1960s, the invention of grounded theory was both a rebuke of sociological methods that assumed the validity of "grand" theorists like Marx, Weber, and Freud and an effort to put forward a more sensible route to generating useful theory. As Christina Goulding (2002) notes, grounded theory is most commonly used "to generate theory where little is already known, or to provide a fresh slant on existing knowledge" (p. 42). The former purpose is invaluable to public engagement with science practitioners, since many engagement events introduce new scientific and technological advancements to publics, and there is little theory by which to analyze their engagement with these new advancements. The latter is also valuable for unseating widely held theories, like the deficit model, which are derived from research with weak ecological validity.

Grounded theory differs from other qualitative research methodologies in several key ways. First, grounded theory's main purpose is theory-building. As Goulding (2002) explains, many qualitative researchers avoid making conceptual links or patterns that could contribute to a transferable theory (p. 44). Beaulieu et al.'s tongue-in-cheek, "Not Another Case

Study," reflects this apprehension. Often qualitative researchers wish to remain in the particularities of specific contexts, and hesitate to begin generating connections with other case studies or theoretical concepts. The results of these hesitancies are made transparent by Geels: the current research milieu is a mess of complex individual case studies, and a proliferation of jargon has led to diminishing returns for many qualitative research studies. Second, an essential feature of grounded theory is its emphasis on parsimony between theory and data (Goulding 2002, p. 45). Here, Glaser and Strauss slightly depart. Although a simplification of their respective positions, Glaser could be said to argue that theory should only explain a phenomenon under study, while Strauss's essential position is that by using a deluge of coding matrices, it is possible to construct theory that validly extends beyond the case at hand (Goulding 2002, p. 45). Insofar as a study attempts to produce ecological validity and utilizes a multi-sited approach (our second middle-range strategy), however, we would argue that theory derived from organic public engagement is applicable across cases.

Since grounded theory's inception, important rifts in the scope and applicability of some core research tenets have developed. As Goulding notes, in nursing, where grounded theory is widely used, researchers are obliged to specify which "brand" of grounded theory they are subscribing to: the original 1967 Glaser and Strauss version, the 1990 Strauss and Corbin edition, or the 1978 or 1992 Glaser approach (Goulding 2002, p. 45). Whole books could be devoted to the differences between these approaches. For our purposes, it will suffice to say that Glaser emphasizes the contextual and emergent aspects of theory construction based on a close adherence to the data, while Strauss stresses a highly complex system of coding techniques that use advanced data manipulation methods to intentionally massage the results of a study. Glaser's 1992 response to Strauss and Corbin's 1990 essay was a scathing condemnation of Strauss's recapitulation of grounded theory, which Glaser saw as completely abandoning many of the principles of parsimony and theory-building that marked their joint venture in the 1960s. Both men had gone down very different career paths, and their respective experiences bifurcate grounded theory into the Glaser and Strauss "camps." Of the two camps, our approach is more Glaserian than Straussarian; Glaser's insistence on parsimony as well as his belief in the theory-building power of grounded theory are more in line with our efforts to produce ecologically valid public engagement methods.

THE VALUE OF HEURISTICS OVER THEORY

Both middle-range theory and grounded theory often produce insights that might better be called heuristics than theories. The value of heuristics and their advantages over larger theories have generated much debate in finance, economics, and psychology over the past twenty years. Scholars and public intellectuals such as Daniel Kahneman, Amos Tversky, Gerd Gigerenzer, and Nassim Taleb have weighed in to various degrees about how simple heuristics guide human decision-making for better and for worse. Where heuristics are especially useful, according to both Gigerenzer and Taleb, is in conditions of uncertainty, where the range of possible outcomes and precise degrees of risk cannot be known. Both scholars frequently compare the domain of theory to statistical thinking, which works well in fixed and controlled conditions, like a craps table, roulette wheel, or controlled laboratory setting. When the variables are controlled, the conditions relatively static, and the patterns fixed, as they are with a pair of dice or a deck of cards, then a total theory of risk and statistical calculation of odds may work quite well.

However, very few existing ecologies reflect these conditions of fixity and control. Whether it be a forest, an ocean, an economy, or a public, actually existing ecologies tend to be prone to unexpected and even unpredictable results, like the trophic cascade caused by the elimination of wolves from Yellowstone we discussed in Chap. 2. This isn't simply a matter of excessive complexity, though that certainly plays a part. Ecologies are also in flux, their interactions, components, and possible outcomes changing over time. Taleb (2007) has famously focused on extremely rare high-impact events, which he calls "black swans." His work has documented how these black swan events calculate to such infinitesimal odds as to be practically impossible, yet such impossible events do and will occur with surprising regularity. John Bogle, financial advisor and founder of the Vanguard investment group, has incorporated Taleb's work into his own paradigm. Bogle has used the October 19, 1987 stock market crash to demonstrate the black swan principle. According to Bogle, by standard methods of risk calculation, the odds of that day's crash were 1 in 10^{50} (which might as well be impossible). In fact, Bogle (2008) argues that 70% of the value of the S&P 500 over 57 years can be accounted for by just 40 days of trading (0.3% of trading days) because these extremely rare days have a such a dramatic impact as to swamp the results of 99.7% of market fluctuations.

Inversely, Gigerenzer tends to focus on more everyday common events and how humans can and do successfully make decisions given that we constantly face conditions of uncertainty. Not only is the vast majority of human decision-making done under such conditions (think just about driving on a busy interstate or having a conversation with a stranger), but Gigerenzer argues that we need incomplete information and heuristics in order to reach conclusions and take action.[1] People need to actively filter out information, even potentially relevant information, so that they can focus on what seems most salient to their decision, which is then made based on a heuristic or even refined intuitions, not grand theories.

In fact, Gigerenzer's research demonstrates that the majority of grand theories and complex parameters ostensibly used to make a decision are either generated post-hoc or modified in the process of application to conform to the results of an intuition or heuristic. The trouble with grand theories is not only that they mask the more productive and valid heuristics in play, but they can start to drive decision-making and displace the simpler heuristic that actually produces the better results that led researchers to try to codify the heuristic into a theory in the first place. Unlike the heuristic or refined intuition, the theory becomes calcified and tends not to adapt to differing or changing ecologies. Just as Maslow's hammer robs researchers of their ability to see better methods of conducting research, these theories rob decision-makers of the ability to adapt their decision processes to new conditions. They begin to see the world as stable and predictable, and, as a result, the theories tend to produce results like the 2008 economic collapse.

Taleb's research on market fluctuations and trading behaviors, along with Gigerenzer's work on heuristics, is having a transformative effect on investing and finance. Daniela Kolusheva compared the performance of sophisticated investing programs with common simple heuristics. Her conclusions were that sophisticated investing rules outperform heuristics only in scenarios where there are no crashes, that is to say, no rare unpredicted significant negative events. In scenarios that included a crash, the heuristics outperform the more sophisticated programs. In fact, even under scenarios of no crashes, the heuristic performed nearly as well as the complex program, losing out only slightly. Meanwhile, in scenarios that included a crash, the advantage of the heuristic over the sophisticated program was far more significant. Unless you believe there is zero future risk of a market crash, investors are probably better off with a simple heuristic than a complex formula or economic theory. The value of heuristics over complex models is also represented by simple index funds outperforming almost all actively managed funds. Likely never having read Taleb or Gigerenzer, but influenced by people like Bogle, retail investors with 401k and IRA plans have increasingly recognized that sophisticated investment programs and expensive professional stock pickers are less successful than a simple heuristic and they have been fleeing actively managed funds in favor of index funds in droves.

We have focused here on the value of heuristics in economics and finance primarily because they are domains of human action that generate enormous amounts of data and some of the most sophisticated and data-driven theories in the world. Economics has, in fact, often become the model for social scientific research, and yet we find that rational economic theory is often outperformed by simple heuristics. Perhaps no other study of human behavior has as much a claim to scientific rationality and analytic precision as economics, and it well displays the fundamental flaws with research that ignores ecological validity and valorizes grand theories over the more useful and often more accurate and predictive middle-range theories and heuristics. Yet, we should emphasize that the value of heuristics is not limited to economics; they are powerful in nearly every domain of human action. As Gigerenzer and Gaissmaier (2011) write, "studies on decisions by individuals and institutions, including business, medical, and legal decisions making, [...] show that heuristics can often be more accurate than complex 'rational' strategies" (p. 473).

When public engagement researchers aim for grand theories of publicness or discourse or develop totalizing models of "the public," odds are good that what we lose are the more accurate and useful heuristics or middle-range theories that can actually inform action in a dynamic way by retaining an attentiveness to local conditions. No doubt good research can refine our intuitions and even improve our heuristics, but only if it surrenders both the rationalist dream that research design and subjects fit into neatly measurable (ideally quantifiable) controlled categories and that the goal of research is to produce some complete and grand theory. Recognizing the value of middle-range theory and the power of heuristics is to accept that sometimes the rule-of-thumb must be not to overthink the situation. Heuristics likewise value the varied and dynamic nature of situations, acknowledging that there may be, in fact, no sufficient model of public deliberation that can best account for or guide all publics, but instead that we must be attentive to each public, each ecology, and each site for its own rules, logics, and methods of operation.

MULTI-SITED ETHNOGRAPHY

Organic public engagement is an iteration of what Marcus calls a "multi-sited ethnography." As he explains, multi-sited ethnography "moves out from the single situations of conventional ethnographic research designs to examine the circulation of cultural meanings, objects, and identities in diffuse time-space" (Marcus 1995, p. 96). While this form of ethnography relies on some macrotheoretical concepts, it does not presume subjects occupy a pre-given contextual position. Given these middle-range features, multi-sited ethnography is a natural fit for producing more ecologically valid research. For example, Hine (2007) notes how multi-sited approaches often "feel necessary in many circumstances as a faithful reflection of lives lived not in discrete locations, but through various forms of connection and circulation." (p. 656). While more traditional ethnographic methods lend practical validity to anthropological research, multi-sited ethnography fulfills these same practical purposes, but does so in an ecologically valid way that can generate middle-range theories by attending to the complexity and itineration of cultural forms. Furthermore, focusing on how these forms circulate in different contexts reflects the complexity of Warner's characterization of publics, which is influenced by the movement, circulation, and distribution of discourses.

Multi-sited ethnography has been an increasingly popular methodology for STS in recent decades, thus setting some precedent for its application to public engagement with science. Early research in STS focused on performing ethnographic studies of laboratories. However, these studies often encountered the same problems that face traditional single-sited ethnographies; mainly, that being deeply entrenched in a single research site increasingly risks compromising and distorting the collected data, or bringing one's own cultural biases to bear on the data, thus skewing analysis. In addition, because academic anthropologists sometimes value what they perceive as exotic, foreign, or other, even to the point of fetishizing unfamiliar cultures, domestic ethnographic work is sometimes undervalued. Marcus (1998) noted how working in one's own culture leads to a "second-class professional citizenship" for many anthropologists, and new researchers were encouraged to do fieldwork among non-English speakers (p. 15). More recently, the realization that individuals participate in multiple publics has lent some credence to reflexive methodologies like auto-ethnographies. Here too, however, ethnography can risk "becoming mere self-quests," vulnerable to "the charges that have so frequently been laid against it in the interest of discrediting reflexive styles of analysis altogether" (Marcus 1998, p. 15). To some extent, multi-sited ethnography sidesteps these problems by examining multiple locations, publics, and discourses, thus preventing the researcher from being inculcated by any given public ecology. The inverse holds true as well; by moving from site-to-site, the risks of interfering with the makeup of the ecology are mitigated. As an example of this shift in STS, Latour and Woolgar's classic laboratory study in *Laboratory Life* differs from Latour's later eclectic *Science and Action* because of his realization that science occurs both inside and outside of the lab. The only way to capture "science in action" is to trace its development through multiple sites.

This itineration has some drawbacks, however. Researchers must adeptly trace complex ideas across numerous sites, and compare and contrast the differences between these public ecologies. Alternatively, many early STS laboratory studies argued for validity by means of their sophisticated understanding of a single site, a feature not found in multi-sited methods. However, as Hine (2007) has argued, this emphasis on single sites offers "problems if taken as a prescription for ongoing work and as a boundary for the legitimate interests of a sociology of science" (p. 659). Part of the problem with deriving research validity from single sites is the tendency to downplay the complexity of publics. Individuals participate in multiple publics simultaneously, and these publics are in part constituted by the environment in which they discourse, deliberate, and debate. Most single-sited engagement events do not capture this facet of publics; since these events are often artificially staged in a single location chosen by the organizers, participants are "prepped" for "proper" deliberation, or divisions between experts and nonexperts are not open for negotiation.

One of the more important upsides of multi-sited ethnography is its potential to resist over-theorization, or theory that artificially dictates the importance of certain kinds of behaviors over others. As we argued above and in Chaps. 4 and 5, establishing artificial conditions or fitting publics into neat categories tends to weaken the ecological validity of public engagement events. Events that lack ecological validity often rely on "naturalized" views of what constitutes publics or expertise, and thus prematurely make sweeping conclusions about various publics' attitudes on the basis of distorted data. Marcus (1998) notes, for instance, how many cultural theories focus "upon the complex construction of subjectivities with particular, but often caricatured, social milieus in mind" (p. 19). Many single-sited engagement events have a parallel problem, mainly that their view of publics are often distilled into a "caricature" called "the public." Multi-sitedness, in contrast, "tends to challenge and complicate in a positive way this hyperemphasis on situated subject positions by juxtaposition and dispersion through investigation in more complex social spaces," and in turn, generates more ecologically valid research. The temptation to prefigure "proper" deliberation is less pervasive if the sites, modes of deliberation, and publics analyzed differ. The more variance between sites, the more likely researchers are to need methodologies born from the distinct circumstances of each ecology. The more that methodologies are born from these distinct circumstances, the more ecologically valid theories derived from the resultant data will be.

USING MULTI-SITED ENGAGEMENT TO BUILD THEORIES AND HEURISTICS

One of the easiest ways to ensure that engagement methodologies reflect the local conditions of each public is to let that public invent, organize, control, and run the engagement event with as little input or interference from the researchers as possible. Selecting a diverse range of sites likewise may help in building robust middle-range theory and useful heuristics, though the nature of that diversity may differ greatly depending on the nature of the research questions and objectives. For example, in Gehrke's research on nanotechnology discussed in Chap. 6, the diversity of publics engaged was based primarily on geography (urban, suburban, and rural publics spread across the country) and on the nature of that location's connection to nanotechnology (academic, government, industry). It could have been figured differently, such as along demographic lines such as race, ethnicity, and sex, which likely would have changed the study significantly. Figuring the diversity of sites differently might well change the codes that emerged, the details behind those codes, and the results of the research.

In the case of Carol Cohn's (2006) work on techno-nuclear discourse and national security discourses, her sites were in part affected by issues of access, particularly when studying military sites and official channels of national security discourse. Not only did she study military bases and government offices, but she observed conferences and meetings, as well as conducted textual analysis. Access issues also led her to use snowball sampling, where one research participant helps to recruit additional participants, and so forth. Snowball sampling can produce issues with biased samples (as participants tend to attract similar participants), but also can help overcome hesitation and distrust.

Rebecca Altman (2008) took a different approach to sampling in her multi-sited ethnography of how citizens and activists use biomonitoring science to track and map chemicals as "they move from the cradles of industrial production to their metaphoric graves in circumpolar communities" (p. 2). Her sampling was based on the three major stages that chemicals follow as they move through this cycle and the different communities where these cycles exist. Hence, her publics were first those where the chemicals were produced, then those that existed where the chemicals were consumed, and finally where the chemicals came to rest (which she calls sites of persistence). In each kind of locale, she finds clusters of stakeholders, whose ways of relating to the chemicals and each other were often conditioned by their position in those chemicals' life cycle. By studying existing community and advocacy groups and organizations in Appalachia's mid-Ohio Valley, Maine, and Alaska, she tracked the chemicals and the ways in which these different sites encountered and deployed biomonitoring.

How one chooses the sites one studies is, thus, a result of the assumptions and designs of the researchers. Ideally the researchers are aware and reflect upon these assumptions and designs, choosing to engage publics that can make the research more robust and useful. Practical considerations may also guide the selection of sites. Access is a common problem with many publics, especially those who view themselves as marginalized, persecuted, or at risk. Some research may choose a very narrow range of publics because the objective of the research is to engage and understand that narrow band. Other research may attempt to reach as broad a diversity of different publics as possible, perhaps to compare between them or to explore what common themes or practices might be present. All of these can be valid approaches, if they are made deliberately and fit the research objectives.

The real challenge of multi-sited ethnography as a guide for public engagement lies not in the selection of the sites but in how one approaches comparison and analysis across the publics being studied. Because so much public engagement research tends to be either idealist or cynical, researchers may be tempted to valorize or demonize specific publics in the study. The assumptions and biases of the researcher can easily lead to comparisons that suggest one public ought to look more or less like another, or that the modes of reasoning and vernacular discourses found in one public are better or worse than the others. Multi-sited ethnography should at least delay such normative judgments, if not forestall them altogether.

Rather than trying to pick out the best apple of the bunch, multi-sited ethnography lets us recognize commonalities that may exceed specific publics under study and point to themes or practices found in a wide range of publics. These can then translate into heuristics, simple guidelines for what might work well or poorly in communicating with a wide variety of publics. Likewise, multi-sited ethnography can reveal the divergences between publics, sometimes stark differences that were unexpected or at least not predictable prior to the event. These also can produce middle-range theories about variations in publics and heuristics that can guide science communication.

Consider, for example, the problem of public knowledge of nanotechnology. Gehrke reports that in two public engagement events the experts who the public group had invited to speak with them were caught off guard by the knowledge of the public group. In one, an adult education lecture program at a local community college in a rural part of a southern state turned out to be inhabited by a number of retired scientists who grilled the invited experts on topics like direction of spin, decay of charge, and specific manufacturing techniques. This might seem predictable in hindsight. The invited speakers could have done a better job of asking the organizers about their audience. They also might have guessed they would encounter some retired scientists, since the town was not far from a major scientific research facility. In the other case, a civic association in a suburban mid-Atlantic town turned out to have a few computer hobbyists sufficiently versed in semiconductor research to throw their invited speaker for a loop (Gehrke, forthcoming). The principle or middle-range theory derived from such experiences may be simply that publics often display unexpectedly robust knowledge of even obscure topics. The heuristics one might generate from these experiences are to always be prepared for an audience that knows more than you do or at least knows much more than you might expect. None of these are earth-shattering realizations, but they are a principle and two heuristics that would serve science communicators well. By using a multi-sited ethnography approach, Gehrke was able to note the repetition of this principle and the value of its corresponding heuristics, which may have been lost in a single event (especially if that event had been organized using conventional public engagement methods).

Building middle-range theories and heuristics is an interpretive and sometimes intuitive approach, as Gigerenzer's research has demonstrated. Through organic public engagement with science, researchers can be guided toward middle-range theories and the production of heuristics by open coding and examining repeated codes or clusters of codes. Yet, when actually formulating the theory and heuristic, researchers will need to observe the patterns and look beyond the simple codes or code clusters to see why and how they repeated. For example, in Chap. 6 we discussed Gehrke's findings that audiences reacted negatively to claims that nanotechnology held promise for cancer treatment. While the repetition of the correlation of "negative response" codes with the topic code for "cancer research" drew attention to the issue, only by watching the videos closely and seeing the reasons, stories, and explanations in each occurrence did the resulting explanation emerge: audiences are skeptical of claims to cancer treatment breakthroughs because so many past claims have over-promised and under-delivered. Now perhaps we can build a middle-range theory: past technologies' failures can affect audience reception of the promise of a new technology, even if the two technologies are unrelated. We may need further research or experience to build a robust heuristic, but we at least now know that speakers should be cautious about making grand promises that a technology will solve wicked problems. That may seem intuitive, but most sound heuristics seem commonsensical after they are actually articulated. Meanwhile, almost all expert speakers in Gehrke's study did not employ this heuristic and tended to make such promises about nanotechnology.

In Cohn's work on techno-nuclear discourse and national security discourses, one of her early insights was a result of how impossible it was for her to engage some of the publics she was studying in a way that revealed their behaviors as they normally existed in their day-to-day practices. As a young, white, female researcher studying communities dominated by competitive masculine culture, she was immediately made aware of how significantly gender norms affected nuclear and security discourses. It was, she reflects, simply impossible that the military personnel might ever speak to her the same way they would speak to one of their buddies, but it was precisely how they performed their roles with her and modified their behavior in her presence that became revelatory about their existing and established group norms. Combining that insight with her range of participant-observer ethnographic engagements, conferences, and meetings, her textual analysis eventually revealed what she calls the "juxtaposition and layering" of the sites (Cohn 2006, p. 107). This is the moment in which something like a middle-range theory of a heuristic emerges, in this case, a heuristic about the role of masculinity in figuring and limiting national security discourse.

Altman's conclusions might similarly be described as a result of juxtaposition and layering. She notes that discourses in sites of production, where regulatory control tends to be well defined and organized, tend to be significantly different than those in sites of consumption and persistence, where regulatory science is less defined. She notes that without the diversity of sites, a study would build a limited heuristic. Studying only sites of production would, she concludes, lead one to believe that biomonitoring is politically contested and "drives a wedge between communities and economic interests" (Altman 2008, p. 383). While studying only sites of persistence might lead one to believe that biomonitoring "serves as an opportunity for cooperation among business, government, and advocacy groups" (Altman 2008, p. 383). By instead using a multi-sited ethnographic approach that traces the lifecycle of the chemicals, she produces a more robust heuristic that recognizes the political importance of biomonitoring and its role in environmental activism, but also notes that it does not, by itself, have a specific political valence. Instead, biomonitoring's role in activism or political contests over chemical manufacture, distribution, and cleanup, is a result of the dynamics of that local community. The middle-range theory here states that biomonitoring is a potent and politically charged element of environmental discourse and regulation, the meaning and role of which will largely be determined by the conditions of the publics and the sites involved in any particular controversy. That is an important and valuable insight, particularly as it displaces the easier, more comfortable, and fundamentally incorrect assumption that biomonitoring has an intrinsic pro-activist bias.

Any multi-sited ethnography has the possibility to produce such middle-range theories and heuristics. Organic public engagement has the advantage of focusing the capacities of multi-sited ethnography on publics and many of the current and persistent problems faced in science communication, public understanding of science, and science policy. Allowing for the more flexible standards of quasi-ethnography and approaching the research from an ecological foundation empowers organic public engagement to make robust and useful contributions in a new and innovative way.

Conclusion

Multi-sited ethnography provides robust data with higher ecological validity than most other methods of collecting data from public engagement events. Such data is ideal for deployment with open coding as part of a

Grounded Theory methodology. This combination of multi-sited ethnography, open coding, and Grounded Theory is a powerful combination for building middle-range theories. Such theories are exactly the kind of research-based, data-driven insights that are both actionable, and yet still adaptable to local conditions. Rather than the kind of grand theories and idealist philosophies that too often dominate public engagement with science, middle-range theories are closer to heuristics, or loose principles that are flexible enough to account for wide variances between publics and the particulars of a specific situation. That flexibility is intrinsic to the method, as the juxtapositions and layering produced by multi-sited ethnography generate more complex and sophisticated data that prevents the imposition of a single grand theory of publics or of proper engagement and deliberation.

The power of middle-range theories and the heuristics they often represent has been well-documented in a wide variety of fields. Heuristics and middle-range theories are especially powerful when confronted with systems or ecologies that are not strictly predictable or do not conform to fixed and knowable outcomes. They have been repeatedly demonstrated as superior to complex rational theories in some of the most studied and data-driven fields that examine human behavior (such as economics and finance). There is every reason to believe that the goal of communicating any topic, but especially complex scientific and technical topics in the wrangle of public and social life, is far better served by building heuristics from multi-sited ethnographies than by any of the currently dominant modes of artificial public engagement.

This conclusion is also bolstered by the existing multi-sited ethnography research and middle-range theory work in science and technology studies. Altman's work on biomonitoring and Cohn's work on techno-nuclear discourse are two explicit examples, but this book has also documented numerous other researchers and studies in public understanding of science that have served as demonstrations of our conclusions. That public engagement with science has not followed this move in science and technology studies and studies of public understanding of science is understandable given its underlying idealist political ideology and its cynical attitude toward publics and their existing means of deliberation. That combination of idealism and cynicism, along with a limited scope of methodologies for studying publics, has pushed public engagement with science into increasing artificial constructs.

The alternative can be found in the long history of ecological thinking and the current research in ecological validity. Ecological thinking dates back to the eighteenth century and the development of an ecological methodology dates back to the nineteenth and early twentieth. Since at least the 1920s and the development of the Chicago School of Sociology, scholars have repeatedly demonstrated the importance of studying people in their existing environments. For example, Julian Steward's 1950s anthropological work demonstrated the powerful influence of environment on human behavior. Of course, we don't want to fall too far into environmental determinism. Instead, our goal is to consider carefully Harold Garfinkel's admonition that we need to account for all the relevant factors of human ecology that are too often ignored, disregarded, or downplayed in so much of the analytic social scientific research.

Since at least the 1940s and 1950s, ecological validity, especially as articulated by Kurt Lewin, has been arguing that researchers should account for ecology as an interwoven dynamic of human and nonhuman elements. This is in many ways simply a common-sense standard. If we wish to study human behaviors, we must study them in the conditions under which such behaviors actually occur or at least in spaces that replicate the relevant factors of those ecologies as much as possible. Since at least the 1960s, researchers calling for an increase in attention to ecologies and ecological validity have developed countless studies demonstrating the importance of ecology and the influence of environment on everything from jury deliberations, political arguments, public learning, and a host of additional human behaviors

In relationship to public engagement with science, we've developed a fivefold standard for ecological validity. Those five standards are:

First, setting, the physical environment and social and cultural environment in which the studied behavior occurs.

Second, sample, that we sample situations or tasks instead of demographic clusters. We sample a diversity of conditions in which the behavior occurs.

The third and fourth are medium and deliberation. The medium of communication and the formal or informal rules and norms of deliberation are important parts of the ecological validity of public engagement.

Fifth, the perceived consequences: What are the likely or expected outcomes of the engagement event?

Dominant modes of public engagement in science have demonstrated very limited ecological validity, producing generally unreliable outcomes and data. Some, especially media studies, oversimplify publics, reducing them to a set of fixed and preestablished interpretive principles. Some promising areas such as studies of science fairs, museums, science cafés, and digital spaces could be significantly enhanced with greater ecological validity, but further study in these genres of public engagement with science is needed. Yet the dominant form of public engagement in both the research and in the consciousness of policy makers and funding agencies is the deliberative engagement, such as consensus conferences, citizen panels, citizen juries, and deliberative polls.

These kinds of studies have demonstrated very weak ecological validity, eschewing actually existing publics in favor of constructing artificial publics for engagement. They aggregate individuals who have never previously constituted any community or public, or individuals who do not share any communication ecology and who do not cohere as a public. These kinds of public engagement events then further guide or alter the deliberative practices of these people, creating artificial public engagements, engaging artificially constructed publics. They create such artificial publics because the researchers and practitioners of these forms of public engagement perceive actually existing publics as flawed, needing either to be told what to believe or to be taught how to have a real dialogue or deliberation.

As a result, these self-styled public engagement experts create a special expertise and a paternalistic authority for themselves while critiquing the tendency for physical scientists to engage in very similar discourses of expertise and paternalism. Public engagement with science researchers and practitioners rarely examine or even express awareness of these common problems and dominant modes of artificial public engagement, precisely because they do not have robust theories of what constitutes a public, how expertise is formed, or the distinction between publics and experts. When they have appealed to political philosophers, such as John Rawls and Jürgen Habermas, the result has been to vacate actually existing publics of all their particulars and locales.

That is, researchers and practitioners of artificial public engagement strip publics of all their naturalness and displace them from their ecologies. The result is that when public engagement experts and practitioners encounter publics that don't fit their preexisting models for ideal deliberative practice, they discount, silence, and even coerce those publics to change their discursive norms, arguing that doing so is justified because it better conforms to their idealized view of democratic discourse. As a result, these artificial public engagements fail to account for how actually existing publics deal with scientific, political, and social controversy. They fundamentally misrepresent public opinion and public values to policy makers.

Artificial public engagement not only silences the vernacular voices of actually existing publics but then claims to speak for those publics, colonizing their voices with their own political ideologies. While on a large scale this can be a problem of researchers striving to describe publics in general and ignoring publics in particular, the prevalence of this silencing and colonization in public engagement methodology and especially public engagement with science also occurs even while ostensibly studying a particular community or public. Public engagement researchers and practitioners deploying methods of artificial public engagement discipline, guide, or even coerce particular publics to behave and communicate in ways that are alien to their day-to-day discursive practices.

As a result, artificial public engagement with science stupefies publics. It misses their unique expertise and values only those voices who will speak in the tongue of the engagement expert. As Alan Irwin argued, the only way forward past this impasse in citizen/science relations is to renounce the belief that publics are passive, homogenous, or incapable. We have over three decades of studies that prove otherwise, yet dominant modes of public engagement with science present scientific inquiry as consistent, harmonious, and progressive, and position publics as nonscientific with knowledge that is illegitimate, undisciplined, and inconsistent. Regardless of the enormous value of science, which is well-documented by its demonstrable success, the practice of science is far from infallible.

We have decades of research and countless cases in which public, local, and "unscientific knowledge" has been far more useful and accurate than the claims of the "scientific experts." Additionally, the complex relationship between scientific expertise and law demonstrates that sometimes social goods, community norms, and values are simply more important or useful than scientific accuracy or expertise. Multiple court cases demonstrate how scientific expertise is negotiable and balanced with other criteria in a deliberative setting. In actually existing publics, we often find these kinds of negotiations over the criteria for expertise and the relative value of different kinds of expertise, whereas in artificial public engagement with science events, these norms are more commonly presented as established and fixed.

While most public engagement with science and science and technology study scholars have been debating the epistemological and ontological status of expert and nonexpert knowledge, we believe they have lost sight of the more practical problem for public engagement researchers and practitioners: how is the status of expert and nonexpert rhetorically negotiated within actually existing publics? The problem of demarcation is not primarily or most importantly a philosophical problem but is a practical and political problem.

Artificial public engagement with science—be it idealist or cynical—has ignored and even obfuscated this fundamental problem by constraining negotiation, debate, and persuasion. Studying actually existing publics as they encounter and negotiate issues of science and technology in their own ways and in their existing ecologies is essential if we want to develop even a rudimentary understanding of the relationship between publics and science. Even modest claims about public opinion and values, much less more sophisticated claims about how publics come to opinion or process information, can only be useful and accurate to the extent that they are studied organically, that is, through methods attentive to the criteria of ecological validity.

For public engagement with science, this means studying naturally existing publics in their native ecologies instead of plucking people out of their physical, social, and cultural environments. The combination of quasi-ethnography and vernacular rhetoric methods found in organic public engagement is especially well-suited to this task. Yet, unlike traditional ethnography and most vernacular rhetoric research, organic public engagement does actively engage the publics it studies, bringing new phenomena into those publics as well as bringing insights out of those publics. To mitigate an engagement event's disruption of a public's existing ecology, organic public engagement follows a minimum intervention standard and lets publics take control of organizing engagement events.

As we detailed in Chap. 6, six principles help guide organic public engagement researchers in their work:

One, publics as ecologies can be synergistic, displaying both positive and negative synergies.
Two, publics as ecologies can be self-organizing and in fact rarely need outside expert help organizing themselves or their events.
Three, publics as ecologies vary in their resiliency.
Four, publics as ecologies vary in their flexibility.
Five, publics as ecologies can evolve.
Six, publics as ecologies can vary in their porousness.

These six principles can both guide analysis of public engagement with science events and serve as reminders to public engagement with science researchers and practitioners that they ought to be attentive and sensitive to the unique characteristics of each public and its ecology.

This means there can never be one easy, uniform, off-the-shelf model of organic public engagement. Instead, organic public engagement is a methodology that must always be adapted to the local conditions. Thus far, the results from work that aligns with the principles of organic public engagement with science have been powerful and practical, as we have documented throughout Chaps. 6 and 7. While they won't generate the grand universal theories of publics or publics discourse that give justification to artificial public engagement with science, organic public engagement events do offer middle-range theories that can make a real difference to understanding and acting in the world.

Combining the quasi-ethnographic and vernacular rhetoric methods that ground organic public engagement with multi-sited ethnography and grounded theory produces a powerful methodology for building heuristics about publics and science communication that are actually meaningful and useful. Likewise, it ensures the public engagement with science events and the research generated about publics more accurately speak for and with actually existing publics rather than silencing them or forcing them to speak in the voice of idealist and cynical engagement experts. Only an attentiveness to ecological validity can promise that kind of accurate and ethical engagement with publics.

In the end, these should be the goals of all public engagement with science: to bring forth the authentic voices of existing publics and to generate useful insights that lead to actionable theories.

NOTES

1. For an excellent review of this research, see Gerd Gigerenzer and Wolgang Gaissmaier, "Heuristic Decision Making," *Annual Review of Psychology* 62 (2011): 451–482.

REFERENCES

Altman, Rebecca Gasior. 2008. *Chemical Body Burden and Place-Based Struggles for Environmental Health and Justice: A Multi-Site Ethnography of Biomonitoring Science*. Doctoral Dissertation. Retrieved from ProQuest (Publication no. 3335628).

Beaulieu, Anne, Andrea Scharnhorst, and Paul Wouters. 2007. Not Another Case Study: A Middle-Range Interrogation of Ethnographic Case Studies in the Exploration of E-science. *Science, Technology, & Human Values* 32 (6): 672–692.

Bogle, John C. 2008. Black Monday and Black Swans. *Financial Analysts Journal* 64 (2): 30–40.

Cohn, Carol. (2006). Motives and methods: Using multi-sited ethnography to study US national security discourses. In: B. Ackerly & J. Tru (Eds.), *Feminist methodologies for international relations.* Cambridge: Cambridge University Press, 91–107.

Geels, Frank W. 2007. Feelings of Discontent and the Promise of Middle Range Theory for STS. *Science, Technology, & Human Values* 32 (6): 627–651.

Gehrke, Pat J. forthcoming. *Nano-Publics: Communicating Nanotechnology Applications, Risks, & Regulations.* New York: Palgrave.

Gigerenzer, Gerd, and Wolgang Gaissmaier. 2011. Heuristic Decision Making. *Annual Review of Psychology* 62: 451–482.

Glaser, Barney, and Anselm Strauss. 1967. *The Discovery of Grounded Theory.* Piscataway: Aldine Transaction.

Goulding, Christina. 2002. *Grounded Theory: A Practical Guide for Management, Business and Market Researchers.* London: Sage.

Hine, Christine. 2007. Multi-Sited Ethnography as a Middle Range Methodology for Contemporary STS. *Science Technology, & Human Values* 32 (6): 652–671.

Marcus, George E. 1995. Ethnography in/of the World System: The Emergence of Multi-Sited Ethnography. *Annual Review of Anthropology* 24: 95–117.

———. 1998. *Ethnography Through Thick and Thin.* Princeton: Princeton University Press.

Merton, Robert K. 1968. *Social Theory and Social Structure.* New York: The Free Press.

Taleb, Nassim N. 2007. *The Black Swan: The Impact of the Highly Improbable.* New York: Random House.

Wyatt, Sally, and Brian Balmer. 2007. Home on the Range: What and Where Is the Middle in Science and Technology Studies. *Science Technology, & Human Values* 32 (6): 619–626.

BIBLIOGRAPHY

Altman, Rebecca Gasior. 2008. *Chemical Body Burden and Place-Based Struggles for Environmental Health and Justice: A Multi-Site Ethnography of Biomonitoring Science*. Doctoral Dissertation. Retrieved from ProQuest (Publication No. 3335628).

Anderson, Ashley A., Jason Delborne, and Daniel Lee Kleinman. 2012. Information Beyond the Forum: Motivations, Strategies, and Impacts of Citizen Participants Seeking Information During a Consensus Conference. *Public Understanding of Science* 22 (8): 956–970.

Arminen, Ilkka. 2008. Scientific and 'Radical' Ethnomethodology. *Philosophy of the Social Sciences* 38 (2): 167–191.

Asen, Robert. 2005. Discourse Theory of Citizenship. *Quarterly Journal of Speech* 90 (2): 189–211.

Barker, Roger G. 1968. *Ecological Psychology: Concepts and Methods for Studying the Environment of Human Behavior*. Stanford: Stanford University Press.

Barker, Roger G., and Herbert F. Wright. 1951. *One Boy's Day: A Specimen Record of Behavior*. New York: Harper and Brothers.

Beaulieu, Anne, Andrea Scharnhorst, and Paul Wouters. 2007. Not Another Case Study: A Middle-Range Interrogation of Ethnographic Case Studies in the Exploration of E-science. *Science, Technology, & Human Values* 32 (6): 672–692.

Bellipanni, Lawrence J., and James Edward Lilly. 1999. What Have Researchers Been Saying About Science Fairs? *Science and Children* 36 (8): 46–50.

Bogle, John C. 2008. Black Monday and Black Swans. *Financial Analysts Journal* 64 (2): 30–40.

Bornstein, Brian H. 1999. The Ecological Validity of Jury Simulations: Is the Jury Still Out? *Law and Human Behavior* 23 (1): 75–91.

© The Author(s) 2018   171
A.S. Lerner, P.J. Gehrke, *Organic Public Engagement*,
https://doi.org/10.1007/978-3-319-64397-7

Breau, David L., and Brian Brook. 2007. 'Mock' Mock Juries: A Field Experiment on the Ecological Validity of Jury Simulations. *Law and Psychology Review* 31: 77–92.

Brown, Mark B. 2006. Survey Article: Citizen Panels and the Concept of Representation. *Journal of Political Philosophy* 14 (2): 203–225.

Cacciatore, Michael A., Dietram A. Scheufele, and Elizabeth A. Corley. 2014. Another (Methodological) Look at Knowledge Gaps and the Internet's Potential for Closing Them. *Public Understanding of Science* 23 (4): 376–394.

Carcioppolo, Nick, Elena V. Chudnovskaya, Andrea Martinez Gonzalez, and Tyler Stephan. 2016. In-Group Rationalizations of Risk and Indoor Tanning: A Textual Analysis of an Online Forum. *Public Understanding of Science* 25 (5): 627–636.

Cheng, Liu Yu. 2012. Ethnomethodology Reconsidered: The Practical Logic of Social Systems Theory. *Current Sociology* 60 (5): 581–598.

Cicourel, Aaron. 2007. A Personal Retrospective View of Ecological Validity. *Text and Talk* 27 (5–6): 735–752.

Citizen Election Forums. *Jefferson Action*. http://jeffersonaction.org/how-we-work/citizens-election-forums/. Accessed 8 July 2016.

Cobb, Michael D. 2005. Framing Effects on Public Opinion About Nanotechnology. *Science Communication* 27 (2): 221–239.

Cohn, Carol. 2006. Motives and Methods: Using Multi-Sited Ethnography to Study US National Security Discourses. In *Feminist Methodologies in International Relations*, ed. Brocke A. Ackerly, Moria Stern, and Jacqui Tru, 91–107. Cambridge: Cambridge University Press.

Cole, Michael, Lois Hood, and Raymond McDermott. 1997. Concepts of Ecological Validity: Their Differing Implications for Comparative Cognitive Research. In *Mind, Culture, and Activity: Seminal Papers from the Laboratory of Comparative Human Cognition*, ed. Michael Cole, Yrjö Engström, and Olga Vasquez, 49–56. Cambridge: Cambridge University Press.

Collins, H.M., and Robert Evans. 2002. The Third Wave of Science Studies: Studies of Expertise and Experience. *Social Studies of Science* 32 (2): 235–296.

Conquergood, Dwight. 1992. Ethnography, Rhetoric, and Performance. *The Quarterly Journal of Speech* 78 (1): 80–97.

Corsi, Pietro. 1988. *The Age of Lamarck Evolutionary Theories in France 1790–1830*. Berkeley: University of California Press.

de Certeau, Michel. 2002. *The Practice of Everyday Life*. Berkeley: University of California Press.

DeLisi, Matt, and Michael G. Vaughn. 2015. The Vindication of Lamarck? Epigenetics at the Intersection of Law and Mental Health. *Behavioral Sciences and the Law* 33 (5): 607–628.

Dewey, John. 1947. *The Public and Its Problems*. Chicago: Gateway Books.

Dhami, Mandeep K., Relph Herwig, and Ulrich Hoffrage. 2004. The Role of Representative Design in an Ecological Approach to Cognition. *Psychological Bulletin* 130 (6): 959–988.

Dijkstra, Anne M., and Christine R. Critchley. 2016. Nanotechnology in Dutch Science Cafés: Public Risk Perceptions Contextualised. *Public Understanding of Science* 25 (1): 71–87.

Dove, Michael R. 2015. Linnaeus' Study of Swedish Swidden Cultivation: Pioneering Ethnographic Work on the 'Economy of Nature'. *AMBIO* 44 (3): 239–248.

Durant, Darrin. 2008. Accounting for Expertise: Wynne and the Autonomy of the Lay Public Actor'. *Public Understanding of Science* 17 (1): 5–20.

———. 2011. Models of Democracy in Social Studies of Science. *Social Studies of Science* 41 (5): 691–714.

Durant, John, Geoffrey Evans, and Geoffrey Thomas. 1992. Public Understanding of Science in Britain: The Role of Medicine in the Popular Representation of Science. *Public Understanding of Science* 1 (2): 161–182.

Edbauer, Jenny. 2005. Unframing Models of Public Distribution: From Rhetorical Situation to Rhetorical Ecologies. *Rhetoric Society Quarterly* 35 (4): 5–24.

Edmond, Gary. 1998. *Down by Science*: Context and Commitment in the Lay Response to Incriminating Scientific Evidence During a Murder Trial. *Public Understanding of Science* 7 (2): 83–111.

Edmond, Gary, and David Mercer. 1997. Scientific Literacy and the Jury: Reconsidering Jury 'Competence'. *Public Understanding of Science* 6 (4): 329–357.

———. 2006. Anti-social Epistemologies. *Social Studies of Science* 36 (6): 843–853.

Einsiedel, Edna F. 2002. Assessing a Controversial Medical Technology: Canadian Public Consultations on Xenotransplantation. *Public Understanding of Science* 11 (4): 315–331.

Emery, Steven B., Henk A.J. Mulder, and Lynn J. Frewer. 2015. Maximizing the Policy Impacts of Public Engagement: A European Study. *Science, Technology, & Human Values* 40 (3): 421–444.

Epstein, Steven. 2000. Democracy, Expertise, and AIDS Treatment Activism. In *Science, Technology, and Democracy*, ed. Daniel D. Kleinman, 15–32. Albany: Sate University of New York Press.

Ercolini, Gina L. 2016. *Kant's Philosophy of Communication*. Pittsburgh: Duquesne University Press.

Evers, Johan, and Joel D'Silva. 2009. Knowledge Transfer from Citizens' Panels to Regulatory Bodies in the Domain of Nano-Enabled Medical Applications. *Innovation: The European Journal of Social Science Research* 22 (1): 125–142.

Fairhead, James, Melissa Leach, and Mary Small. 2006. Public Engagement with Science? Local Understandings of a Vaccine Trial in the Gambia. *Journal of Biosocial Science* 38 (1): 103–116.

Feinberg, Andrew P., and Bert Vogelstein. 1983. Hypomethylation Distinguishes Genes of Some Human Cancers from Their Normal Counterparts. *Nature* 301 (6): 89–92.

Felt, Ulrike, and Maximillian Fochler. 2008. The Bottom-Up Meanings of the Concept of Public Participation in Science and Technology. *Science and Public Policy* 35 (7): 489–499.

Fishkin, James S., Robert C. Luskin, and Roger Jowell. 2000. Deliberative Polling and Public Consultation. *Parliamentary Affairs* 53 (4): 657–666.

Fiske, John. 1989. *Reading the Popular*. London: Routledge.

Foucault, Michel. 1970. *The Order of Things*. New York: Vintage Books.

Fuller, Steven. 2006a. A Step Toward the Legalization of Science Studies. *Social Studies of Science* 36 (6): 827–834.

———. 2006b. *The Philosophy of Science and Technology Studies*. New York: Routledge.

Garfinkel, Harold. 1967. *Studies in Ethnomethodology*. Englewood Cliffs: Prentice-Hall.

Geels, Frank W. 2007. Feelings of Discontent and the Promise of Middle Range Theory for STS. *Science, Technology, & Human Values* 32 (6): 627–651.

Gehrke, Pat J. 2014. Ecological Validity and the Study of Publics: The Case for Organic Public Engagement Methods. *Public Understanding of Science* 23 (1): 77–91.

———. forthcoming-a. "Ecological Validity." In *The SAGE Encyclopedia of Educational Research, Measurement, and Evaluation*, ed. Bruce Frey. Thousand Oaks: SAGE.

———. forthcoming-b. *Nano-Publics: Communicating Nanotechnology Applications, Risks, & Regulations*. New York: Palgrave.

Gibson, James J. 1979. *The Ecological Approach to Visual Perception*. Boston: Houghton Mifflin Company.

Gigerenzer, Gerd, and Wolgang Gaissmaier. 2011. Heuristic Decision Making. *Annual Review of Psychology* 62: 451–482.

Glaser, Barney, and Anselm Strauss. 1967. *The Discovery of Grounded Theory*. Piscataway: Aldine Transaction.

Goulding, Christina. 2002. *Grounded Theory: A Practical Guide for Management, Business and Market Researchers*. London: Sage.

Goven, Joanna. 2003. Deploying the Consensus Conference in New Zealand: Democracy and De-problematization. *Public Understanding of Science* 12 (4): 423–440.

Graham, S. 2015. *Scott. The Politics of Pain Medicine: A Rhetorical-Ontological Inquiry*. Chicago: University of Chicago Press.

Gross, Alan G. 2006. *Starring the Text: The Place of Rhetoric in Science Studies*. Carbondale: Southern Illinois University Press.

Guttman, Amy, and Dennis Thompson. 1998. *Democracy and Disagreement*. Cambridge: The Belknap Press of Harvard University Press.

Habermas, Jürgen. 1995. Reconciliation Through the Public Use of Reason: Remarks on John Rawls's Political Liberalism. *The Journal of Philosophy* 92 (3): 109–131.

Haeckel, Ernst. 1866. *Generelle Morphologie*. Berlin: Georg Reimer. Quoted in Robert J. Richards. 2008. *The Tragic Sense of Life: Ernst Haeckel and the Struggle Over Evolutionary Thought*, 144. Chicago: University of Chicago Press.

Hammersley, Martyn. 1990. What's Wrong with Ethnography? The Myth of Theoretical Description. *Sociology* 24 (4): 597–615.

Hammond, Kenneth R. 1978. Psychology's Scientific Revolution: Is It in Danger? *Center for Research on Judgment and Policy* 211: 1–45.

Harris, Marvin. 1976. History and Significance of the Emic/Etic Distinction. *Annual Review of Anthropology* 5: 329–350.

Hartley, Troy W., and Robert A. Robertson. 2006. Stakeholder Engagement, Cooperative Fisheries Research and Democratic Science: The Case of the Northeast Consortium. *Human Ecology Review* 13 (2): 161–171.

Hauser, Gerald A. 1999. *Vernacular Voices: The Rhetoric of Publics and Public Spheres*. Columbia: University of South Carolina Press.

———. 2007. Vernacular Discourse and the Epistemic Dimension of Public Opinion. *Communication Theory* 17: 333–339.

Henriksen, Ellen K., and Merethe Frøyland. 2000. The Contribution of Museums to Scientific Literacy: Views from Audience and Museum Professionals. *Public Understanding of Science* 9 (4): 393–415.

Hicks, Darrin. 2002. The Promise(s) of Deliberative Democracy. *Rhetoric & Public Affairs* 5 (2): 223–260.

Hine, Christine. 2007. Multi-Sited Ethnography as a Middle Range Methodology for Contemporary STS. *Science, Technology, & Human Values* 32 (6): 652–671.

Hippocrates. On the Sacred Disease. *The Internet Classics Archive*. http://classics.mit.edu/Hippocrates/sacred.html.

Howard, Robert Glenn. 2010. The Vernacular Mode: Locating the Non-intuitional in the Practice of Citizenship. In *Public Modalities: Rhetoric, Culture, Media, and the Shape of Public Life*, ed. Daniel C. Brouwer, 240–261. Tuscaloosa: University of Alabama Press.

Hutton, James. 1788. *Theory of the Earth*. Edinburgh: The Royal Society of Edinburgh.

Irwin, Alan. 1995. *Citizen Science: A Study of People and Expertise and Sustainable Development*. New York: Routledge.

Jasanoff, Sheila. 1995. *Science at the Bar*. Cambridge: Harvard University Press.

———. 2003. Breaking Waves in Science Studies: Common on H.M. Collins and Robert Evans, 'The Third Wave of Science Studies'. *Social Studies of Science* 33 (3): 389–400.

Jensen, Eric, and Nicola Buckley. 2014. Why People Attend Science Festivals: Interests, Motivations and Self-Reported Benefits of Public Engagement with Research. *Public Understanding of Science* 23 (5): 557–573.

Jones, Richard. 2007. What Have We Learned from Public Engagement? *Nature Nanotechnology* 2 (5): 262–263.

Jones, Blake L., and David Royse. 2008. Citizen Review Panels for Child Protective Services: A National Profile. *Child Welfare* 87 (3): 142–162.

Judkins, Gabriel, Marissa Smith, and Eric Keys. 2008. Determinism Within Human-Environment Research and the Rediscovery of Environment Causation. *The Geographical Journal* 174 (1): 17–29.

Kamolpattana, Supara, Ganigar Chen, Pichai Sonchaeng, Clare Wilkinson, Neil Willey, and Karen Bultitude. 2015. Thai Visitors' Expectations and Experiences of Explainer Interaction Within a Science Museum Context. *Public Understanding of Science* 24 (1): 69–85.

Kant, Immanuel. 2006. An Answer to the Question: What Is Enlightenment? In *Toward Perpetual Peace and Other Writings on Politics, Peace, and History*, ed. Pauline Kleingeld, trans. David L Colclasure. New Haven: Yale University Press.

Kaplan, Abraham. 1964. *The Conduct of Inquiry: Methodology for Behavioral Science*. Rutgers: Transaction Publishers.

King, Gerry, David J. Heaney, David Boddy, Catherine A. O'Donnell, Julia S. Clark, and Frances S. Mair. 2011. Exploring Public Perspectives on E-health: Findings from Two Citizen Juries. *Health Expectations* 14 (4): 351–360.

Kleinman, Daniel. 2000. Democratization of Science and Technology. In *Science, Technology, and Democracy*, ed. Daniel Kleinman, 139–165. Albany: State University of New York Press.

Koerner, Lisbet. 1999. *Linnaeus: Nature and Nation*. Cambridge: Harvard University Press.

Kuhn, Thomas. 2012. *The Structure of Scientific Revolutions*. 50th anniversary ed. Chicago: University of Chicago Press.

Ladouceur, Robert, Anne Gaboury, Annie Bujold, Nadine Lachance, and Sarah Tremblay. 1991. Ecological Validity of Laboratory Studies of Videopoker Gaming. *Journal of Gambling Studies* 7 (2): 109–116.

Laudan, Larry. 1983. The Demise of the Demarcation Problem. In *Physics, Philosophy and Psychoanalysis*, ed. R.S. Cohen, and L. Laudan, 111–127. Dordrecht: D. Reidel Publishing Company.

Lave, Jean. 1997. What's Special About Experiments as Contexts for Thinking. In *Mind, Culture, and Activity: Seminal Papers from the Laboratory of Comparative Human Cognition*, ed. Michael Cole, Yrjö Engström, and Olga Vasquez, 57–69. Cambridge: Cambridge University Press.

Lee, Caroline W. 2016. *Do-It-Yourself Democracy*. New York: Oxford University Press.

Lévy-Leblond, Jean-Marc. 1992. About Misunderstandings About Misunderstandings. *Public Understanding of Science* 1 (1): 17–21.

Lewin, Kurt. 1943. Defining the 'Field at a Given Time'. *Psychological Review* 50 (3): 292–310.

Lippmann, Walter. 1927. *The Phantom Public*. New Brunswick: The Macmillan Company.

Lloyd, G.E.R. 1979. *Magic, Reason and Experience: Studies in the Origin and Development of Greek Science*. Cambridge: Cambridge University Press.

———. 2009. *Disciplines in the Making: Cross-Cultural Perspectives on Elites, Learning, and Innovation*. Oxford: Oxford University Press.

Locke, Simon. 1999. Golem Science and the Public Understanding of Science: From Deficit to Dilemma. *Public Understanding of Science* 8 (2): 75–92.

Lynch, Michael. 2006. From Ruse to Farce. *Social Studies of Science* 36 (6): 819–826.

Lynch, Michael, and Simon Cole. 2005. Science and Technology Studies on Trial: Dilemmas of Expertise. *Social Studies of Science* 35 (2): 269–311.

Lynch, Paul, and Nathaniel Rivers, eds. 2015. *Thinking with Bruno Latour in Rhetoric and Composition*. Carbondale: Southern Illinois University Press.

MacDonald, Dennis W. 2011. Beyond the Group: The Implications of Roderick D. McKenzie's Human Ecology for Reconceptualizing Society and the Social. *Nature and Culture* 6 (3): 268–284.

Macnaghten, Phil, Matthew Kearns, and Brian Wynne. 2005. Nanotechnology, Governance and Public Deliberation: What Role for the Social Sciences? *Science Communication* 27 (2): 268–291.

Macoubrie, Jane. 2006. Nanotechnology: Public Concerns, Reasoning and Trust in Government. *Public Understanding of Science* 15 (2): 221–241.

Mair, Michael, Christian Greiffenhagen, and W.W. Sharrock. 2016. Statistical Practice: Putting Society on Display. *Theory, Culture & Society* 33 (3): 51–77.

Marcus, George E. 1995. Ethnography in/of the World System: The Emergence of Multi-Sited Ethnography. *Annual Review of Anthropology* 24: 95–117.

Maslow, Abraham. 1966. *Psychology of Science: A Reconnaissance*. New York: Harper & Row.

Master, Karen, Eun Young, Joe Cox, Brooke Simmons, Chris Lintott, Gary Graham, Anita Greenhill, and Kate Holmes. 2016. Science Learning Via Participation in Online Citizen Science. *Journal of Science Communication* 15 (3): 1–33.

McCormick, Samuel. 2003. Earning One's Inheritance: Rhetorical Criticism, Everyday Talk, and the Analysis of Public Discourse. *Quarterly Journal of Speech* 89 (2): 109–131.

McKenzie, Roderick D. 1924. The Ecological Approach to the Study of the Human Community. *American Journal of Sociology* 30 (3): 287–301.

Measles Cases and Outbreaks. *Centers for Disease Control and Prevention*. http://www.cdc.gov/measles/cases-outbreaks.html.

Menon, Devidas, and Tania Stafinski. 2008. Engaging the Public in Priority-Setting for Health Technology Assessment: Findings from a Citizen's Jury. *Health Expectations* 11 (3): 282–293.

Mercer, David. 2016. Why Popper Can't Resolve the Debate Over Global Warming: Problems with the Uses of Philosophy of Science in the Media and Public Framing of the Science of Global Warming. *Public Understanding of Science*: 1–14.

Merkle, Daniel M. 1996. The Polls—Review: The National Issues Convention Deliberative Poll. *Public Opinion Quarterly* 60 (4): 588–619.

Merton, Robert K. 1968. *Social Theory and Social Structure*. New York: The Free Press.

Michael, Mike. 2009. Publics Performing Publics: Of PiGs, PiPs and Politics. *Public Understanding of Science* 18 (5): 617–631.

Miller, Jon D. 1998. The Measurement of Civic Scientific Literacy. *Public Understanding of Science* 7 (1): 203–223.

Min, Seong-Jae. 2007. Online vs. Face-to-Face Deliberation: Effects on Civic Engagement. *Journal of Computer-Mediated Communication* 12 (4): 1369–1387.

Mol, Annemarie. 2002. *The Body Multiple: Ontology in Medical Practice*. Durham: Duke University Press.

Mouffe, Chantal. 1999. Deliberative Democracy or Agonist Pluralism? *Social Research* 66 (3): 745–758.

———. 2009. The Limits of Jon Rawls' Pluralism. *Theoria: A Journal of Social and Political Theory* 56 (118): 1–14.

Munno, Greg, and Tina Nabatchi. 2014. Public Deliberation and Co-production in the Political and Electoral Arena: A Citizens' Jury Approach. *Journal of Public Deliberation* 10 (2): 1–29.

Nisbet, Erik C. 2006. The Engagement Model of Opinion Leadership: Testing Validity Within a European Context. *International Journal of Public Opinion Research* 18 (1): 3–30.

O'Doherty, Kieran, and Alice Hawkins. 2010. Structuring Public Engagement for Effective Input in Policy Development on Human Tissue Biobanking. *Public Health Genomics* 13 (4): 197–206.

Ono, Kent A., and John M. Sloop. 1995. The Critique of Vernacular Discourse. *Communication Monographs* 62 (1): 19–46.

Orlove, Benjamin S. 1980. Ecological Anthropology. *Annual Review of Anthropology* 9 (23): 235–273.

Pardo, Rafael, and Félix Calvo. 2002. Attitudes Toward Science Among the European Public: A Methodological Analysis. *Public Understanding of Science* 11 (2): 155–195.

Park, Robert E., Ernest W. Burgess, and Roderick D. McKenzie. 1925. *The City*. Chicago: University of Chicago Press.

Pidgeon, Nick, and Tee Rogers-Hayden. 2007. Opening Up Nanotechnology Dialogue with the Publics: Risk Communication of 'Upstream Engagement'. *Health, Risk & Society* 9 (2): 191–210.

Pollner, Melvin. 2012. The End(s) of Ethnomethodology. *American Sociology* 43 (1): 7–20.

Popper, Karl. 1992. *The Logic of Scientific Discovery*. New York: Routledge.

Powell, Maria C., and Mathilde Colin. 2009. Participatory Paradoxes: Facilitating Citizen Engagement in Science and Technology from the Top-Down? *Bulletin of Science, Technology, and Society* 29 (4): 325–342.

Ranger, Mathieu, and Karen Bultitude. 2016. 'The Kind of Mildly Curious Sort of Science Interested Person Like Me': Science Bloggers' Practices Relating to Audience Recruitment. *Public Understanding of Science* 25 (3): 361–378.

Rawls, John. 1993. *Political Liberalism*. New York: Columbia University Press.

Reynolds, Andrew. 2008. Ernst Haeckel and the Theory of the Cell State: Remarks on the History of a Bio-political Metaphor. *History of Science* 46 (2): 123–152.

Richards, Robert J. 2008. *The Tragic Sense of Life: Ernst Haeckel and the Struggle Over Evolutionary Thought*. Chicago: Chicago University Press.

Rickert, Thomas. 2013. *Ambient Rhetoric: The Attunements of Rhetorical Being*. Pittsburgh: University of Pittsburgh Press.

Rowe, Gene, and Lynn J. Frewer. 2005. A Typology of Public Engagement Mechanisms. *Science, Technology, & Human Values* 30 (2): 251–290.

Schiele, Bernard. 2008. Science Museums and Science Centres. In *Handbook of Public Communication of Science and Technology*, ed. Massimiano Bucchi and Brian Trench, 27–39. New York: Routledge.

Schweigert, Francis J. 2010. Strengthening Citizenship Through Deliberative Polling. *Journal of Community Practice* 18 (1): 19–39.

Shein, Paichi P., Yuh-Yuh Li, and Tai-Chi Huang. 2015. The Four Cultures: Public Engagement with Science Only, Art Only, Neither, or Both Museums. *Public Understanding of Science* 24 (8): 943–956.

Simis, Molly J., Haley Madden, Michael A. Cacciatore, and Sara K. Yeo. 2016. The Lure of Rationality: Why Does the Deficit Model Persist in Science Communication? *Public Understanding of Science* 25 (4): 400–414.

Sloop, John, and Kent Ono. 1997. Out-Law Discourse: The Critical Politics of Material Judgment. *Philosophy and Rhetoric* 30 (1): 51–69.

Steward, Julian H. 1972. *Theory of Culture Change: The Methodology of Multilinear Evolution*. Champaign: University of Illinois Press.

Stilgoe, Jack. 2007. *Nanodialogues: Experiments in Public Engagement with Science*. London: Demos.

Stocking, S. Holly, and Lisa W. Holstein. 2009. Manufacturing Doubt: Journalists' Roles and the Construction of Ignorance in a Scientific Controversy. *Public Understanding of Science* 18 (1): 24–42.

Stone, John V. 2002. *Public Participation in Environmental Management: Seeking Participatory Equity Through Ethnographic Inquiry*. PhD Dissertation, University of South Florida.

Street, Jackie, Katherine Duszynski, Stephanie Krawczyk, and Annette Braunack-Mayer. 2014. The Use of Citizens' Juries in Health Policy Decision-Making: A Systematic Review. *Social Science and Medicine* 109: 1–9.

Szokolszky, Agnes. 2013. Interview with Ulric Neisser. *Ecological Psychology* 25 (2): 182–199.

Taleb, Nassim N. 2007. *The Black Swan: The Impact of the Highly Improbable*. New York: Random House.

Trench, Brian. 2008. Internet: Turning Science Communication Inside-Out? In *Handbook of Public Communication of Science and Technology*, ed. Massimiano Bucchi and Brian Trench, 185–198. New York: Routledge.

Trumbo, Craig. 1996. Constructing Climate Change: Claims and Frames in US News Coverage of an Environmental Issue. *Public Understanding of Science* 5 (3): 269–283.

Warner, Michael. 2005. *Public and Counterpublics*. New York: Zone Books.

Wilkinson, Clare Stuart Allan, Alison Anderson, and Alan Petersen. 2007. From Uncertainty to Risk?: Scientific and News Media Portrayals of Nanoparticle Safety. *Health, Risk & Society* 9 (2): 145–157.

Williamson, Julie R., and Daniel Sundén. 2015. Deep Cover HCI: A Case for Covert Research in HCI. In *CHI EA 15: 33rd Annual ACM Conference Extended Abstracts on Human Factors in Computing Systems April 18–23*, 543–554.

Wulf, Andrea. 2015. *The Invention of Nature: Alexander Von Humboldt's New World*. New York: Alfred A. Knopf.

Wyatt, Sally, and Brian Balmer. 2007. Home on the Range: What and Where Is the Middle in Science and Technology Studies? *Science, Technology, & Human Values* 32 (6): 619–626.

Wynne, Brian. 1991. Knowledges in Context. *Science, Technology & Human Values* 16 (1): 111–121.

———. 1992a. Public Understanding of Science Research: New Horizons or Hall of Mirrors? *Public Understanding of Science* 1 (1): 37–43.

———. 1992b. Misunderstood Misunderstanding: Social Identities and Public Uptake of Science. *Public Understanding of Science* 1 (3): 283–304.

———. 2003. Seasick on the Third Wave? Subverting the Hegemony of Propositionalism. *Social Studies of Science* 33 (3): 401–417.

———. 2008. Elephants in The Rooms Where Publics Encounter 'Science'?: A Response to Darrin Durant, 'Accounting for Expertise: Wynne and the Autonomy of the Lay Public. *Public Understanding of Science* 17 (1): 21–33.

Yaneva, Albena, Tania Mara Rabesandratana, and Birgit Greiner. 2009. Staging Scientific Controversies: A Gallery Test on Science Museums' Interactivity. *Public Understanding of Science* 18 (1): 79–90.

INDEX

A

actor-network theory (ANT), 149, 150
AIDS research, 32, 99
apodictic certainty, 102
Aristotle, 8, 101–3, 116
artificial public engagement, 166–7
Asen, Robert, 71, 86, 87

B

Barker, James G., 18–20
black swan events, 154
Brunswick, Egon, 24
Burgess, Ernest, 14–15

C

Chernobyl, 123, 124, 131
Chicago School, 13–15, 165
citizen juries, 30
citizen panels, 58–60
citizen science, 53
climate change, 104
coding, 144, 153, 162
cognitive psychology, 26–7

computer-assisted qualitative data analysis, 144
consensus conferences, 56–8
constructivism, 108–10, 112, 113
context, 116
counter-factual research, 55–6
cultural ecology, 15–18
cynical engagement, 27, 69–71, 76, 77, 81, 82, 100, 101, 106, 107, 110, 111, 140, 160, 168

D

deficit model, 107, 147
deliberative democratic theory, 77
deliberative engagement methodologies, 53–6
deliberative polls, 62–4
demarcation problem, 98, 100
demographics, 30, 85
determinism, 15, 18, 19
Dewey, John, 70, 74–7
  and idealism, 76
digital spaces, 51–3
DNA testing, 109
downstream engagement, 98, 100

© The Author(s) 2018
A.S. Lerner, P.J. Gehrke, *Organic Public Engagement*,
https://doi.org/10.1007/978-3-319-64397-7

The manufacturer's authorised representative in the EU is Springer
Nature Customer Service Centre GmbH, Europaplatz 3, 69115 Heidelberg,
Germany. If you have any concerns regarding our products, please
contact ProductSafety@springernature.com

Printed and bound by CPI Group (UK) Ltd, Croydon, CR0 4YY
27/04/2026
02097607-0001